Andrew M. Peterson, Jahan Marcu (Eds.)
Demystifying Cannabis and Hemp

Also of interest

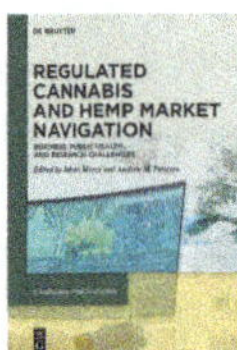

Regulated Cannabis and Hemp Market Navigation.
Business, Public Health, and Research Challenges
Jahan Marcu, Andrew M. Peterson (Eds.), 2024
ISBN 978-3-11-067742-3
e-ISBN 978-3-11-067750-8

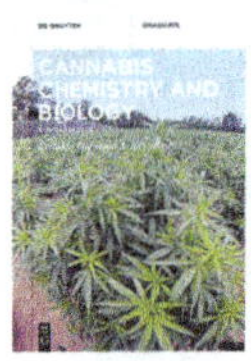

Cannabis Chemistry and Biology.
Fundamentals
Mahmoud A. ElSohly (Ed.), 2023
ISBN 978-3-11-071835-5
e-ISBN 978-3-11-071836-2

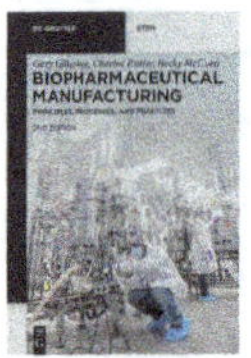

Biopharmaceutical Manufacturing.
Principles, Processes, and Practices
Gary Gilleskie, Charles Rutter and Becky McCuen (Eds.), 2025
ISBN 978-3-11-111206-0
e-ISBN 978-3-11-111245-9

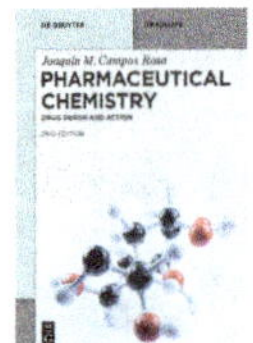

Pharmaceutical Chemistry.
Drug Design and Action
Joaquín M. Campos Rosa, 2024
ISBN 978-3-11-131654-3
e-ISBN 978-3-11-131690-1

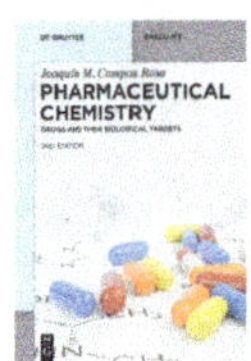

Pharmaceutical Chemistry.
Drugs and Their Biological Targets
Joaquín M. Campos Rosa, 2024
ISBN 978-3-11-131655-0
e-ISBN (PDF) 978-3-11-131688-8

Demystifying Cannabis and Hemp

Myths, Mysteries, and Truths

Edited by
Andrew M. Peterson and Jahan Marcu

 **DE GRUYTER**

Editors
Andrew M. Peterson
623 Kingsley Drive
Ventnor City, NJ 08406
United States of America
apeterson@sju.edu

Jahan Marcu
Delaware Analytical, LLC
1196 S Little Creek Rd
Suite 103
Dover, Delaware 19901
United States of America
Info@DelawareAnalytical.com

ISBN 978-3-11-147513-4
ISBN 978-3-11-147616-2 (PDF)
ISBN 978-3-11-147714-5 (EPUB)
DOI https://doi.org/10.1515/9783111476162

Library of Congress Control Number: 2026937906

Bibliographic information published by the Deutsche Nationalbibliothek
The Deutsche Nationalbibliothek lists this publication in the Deutsche Nationalbibliografie; detailed
bibliographic data are available on the internet at http://dnb.dnb.de.

© 2026 Walter de Gruyter GmbH, Genthiner Straße 13, 10785 Berlin, Germany

De Gruyter and Walter de Gruyter GmbH are part of De Gruyter Brill.
www.degruyterbrill.com

Questions about General Product Safety Regulation:
productsafety@degruyterbrill.com

Cover illustration: boonstudio/iStock/Getty Images Plus

Preface: The Plant That Refuses to Behave

Cannabis is a plant that refuses to behave. It slips easily from medicine to menace, from sacrament to Schedule I, from backyard crop to billion-dollar asset, from folk remedy to patented molecule. Each time we believe we have finally decided what *Cannabis* is, history reframes the question and invites us to reconsider. Few subjects sit so uncomfortably at the intersection of biology, law, medicine, commerce, and culture. Fewer still carry such a dense accumulation of fear, propaganda, improvisation, and scientific neglect.

For most of the last century, *Cannabis* was treated as a social problem in search of justification for its punishment. Then, almost overnight, it reemerged as a therapeutic commodity in search of regulatory legitimacy. Along the way it became a proxy for larger struggles – over race and policing, federal power and state rebellion, corporate consolidation and legacy survival, expertise and lived experience. The result is not coherence but a patchwork: of laws written for different fears, of data collected for different purposes, of markets governed by different logics depending on geography. *Cannabis* lives inside this patchwork as both beneficiary and casualty.

This book is organized around three movements that reflect how *Cannabis* itself has been understood across time: **myths, mysteries, and truths**. These are not clean stages that replace one another. They overlap, bleed together, and often masquerade as one another. But they provide a useful compass for navigating a field thick with certainty and thin on consensus.

Chapters 1 through 4 confront the myths. These include botanical myths about species and strains, chemical myths about danger and safety, and cultural myths about prohibition as a rational response to scientific risk. These chapters do not simply debunk false beliefs; they show how such narratives arise, why they persist, and how easily they harden into institutional fact.

Chapters 5 through 8 explore the mysteries. These are the unsolved problems of modern *Cannabis* governance: how stigma outlives prohibition, how global markets fracture under incompatible laws, how data is produced inside regulatory bottlenecks, how paradox multiplies rather than resolves as reform accelerates. These chapters focus on what we *do not yet fully understand*, even as policy, commerce, and medicine move forward with confidence.

Chapters 9 through 12 turn toward what we can responsibly call truths – or, more precisely, toward the most reliable knowledge currently available. Here the reader encounters courts wrestling with expertise, medicine grappling with evidence and risk, and science acknowledging the limits of what it can promise. These truths are provisional by design. They rest not on certainty, but on method.

Across all three movements runs a common tension: the collision between human institutions and biological reality. *Cannabis* remains stubbornly indifferent to our taxonomies, our statutes, and our market projections. It grows according to environmental constraints, expresses genetic tendencies without regard for legal categories, and produces molecules indifferent to moral frameworks. In that sense, the longest argument in this book is not between advocates and opponents, regulators and entrepreneurs, medicine and law. It is between our systems of control and a plant that resists them.

This is not a manifesto for universal legalization, nor a brief for renewed prohibition. It is not a celebration, and it is not an indictment. It is an attempt to take *Cannabis* seriously – scientifically, legally, politically, and ethically – without the shortcuts of moral panic or commercial enthusiasm. Where certainty is unavailable, we say so. Where systems conflict, we describe the conflict rather than resolve it rhetorically.

Readers will come to this book from different positions: clinicians, lawyers, scientists, policymakers, operators, and patients. No single audience owns this subject – just as no single institution owns the plant. If this book succeeds, it will not deliver easy answers. It will offer clearer questions, sharpened by evidence, history, and a sober respect for uncertainty.

And perhaps, along the way, it will deepen our appreciation for a plant that, after centuries of cultivation and a century of fear, still keeps slipping beyond the story we tell about it.

Contents

TRUTHS

List of Contributors

Molly E. Burnett, PharmD
Swedesboro, NJ, USA
Mollburnett@yahoo.com

Chris Conrad
Oaksterdam University
Oakland, CA, USA
case@chrisconrad.com

Ruth Fisher, PhD
Quantaa
St. Petersburg, FL, USA
ruth@quantaa.com

Cheryl J. Fitzer-Attas PhD, MBA, President
ClinMed LLC
Monroe Township, NJ, USA
cheryl@clinmedaffairs.com

Frederick J. Goldstein PhD, FCP
Philadelphia College of Osteopathic Medicine
Philadelphia, PA, USA
fredg@pcom.edu

Scott D. Greene RPh, MS, PhD
Philadelphia, PA, USA
sgreene203@aol.com

Robert T. Hoban, Esq
Cannabis Policy Institute
UNLV
Las Vegas, NV, USA
bob@bobhoban.com

Mariana Larrea-Arias, Esq
Lifesciences and Cannabis Attorney
Mexico City, Mexico
mlarrea@mlawboutique.com

Michael T. Loconto, Esq
Boston, MA, USA
mtl@locontoadr.com

Jahan Marcu, PhD
Delaware Analytical, LLC
Dover, Delaware, DE, USA
info@delawareanalytical.com

Lila McKinley, J.D
Cannabis Control Division
Connecticut Department of Consumer Protection
Hartford, CT, USA
lila.mckinley@ct.gov

Andrew M. Peterson PharmD, PhD, FCPP
Department of Pharmacy
Philadelphia College of Pharmacy/Saint Joseph's
University
Philadelphia, PA, USA
apeterson@sju.edu

Bianca Schnarr, MS
Thomas Jefferson University and
Hudson County Community College
Institute of Emerging Health Professions
School of Business, Culinary Arts, and Hospitality
Management
Philadelphia, PA, USA
schnarrb@gmail.com

Anna Schwabe, PhD
University of Colorado Boulder
New Jersey, CO, USA
annaschwabe@outlook.com

© 2026 Walter de Gruyter GmbH, Berlin | https://doi.org/10.1515/9783111476162-204

MYTHS

Anna Schwabe

Knowing Species Matters: Taxonomy and Naming in Cannabis

Introduction

Understanding plant species is important, though the reasons differ depending on who is asking and in what context. Scientists use species names because they provide a standardized, universally accepted system that avoids confusion. Scientific names, based on Latin binomials, are consistent across languages and regions, allowing researchers in different parts of the world to know exactly which plant is being discussed.

Plant taxonomy is the scientific discipline that identifies, describes, classifies, and names plants based on their morphological (physical) and phylogenetic (evolutionary) relationships. It organizes the diversity of plants into a hierarchical system of increasingly specific categories (Kingdom, Phylum, Class, Order, Family, Genus, Species, and lower taxa such as subspecies, variety, and form), which provides a logical framework for understanding relationships and managing botanical knowledge. Key aspects of plant taxonomy include identification, description, classification, and nomenclature in accordance with the International Code of Nomenclature (ICN) for algae, fungi, and plants [1].

The public, on the other hand, usually relies on common names, which are easier to say and remember. While practical, this system comes with challenges. A single plant can have multiple common names, and one common name can refer to different plants depending on the region. A bluebell in the United Kingdom refers to *Hyacinthoides non-scripta* in the Asparagaceae family, but in the United States, a bluebell refers to *Mertensia virginica* in the Boraginaceae family. These mismatches can cause confusion in gardening, herbal medicine, or foraging, where precision and accuracy are important. Common names may also give the wrong impression about a plant's identity. A pineapple, despite its name, is neither a pine nor an apple. Similarly, peanuts are not true nuts, but legumes. These examples highlight how everyday naming often reflects appearance, taste, or cultural history, rather than botanical reality.

In some cases, misunderstanding plant identification can have serious consequences. Confusing *Daucus carota* commonly referred to as wild carrot or "Queen Anne's lace" with *Conium maculatum* commonly known as poison hemlock when foraging can lead to death. In agriculture and commerce, clarity in species identification affects trade, labeling, and consumer trust. The same is true in *Cannabis sativa* L., which has many common names [2], refers to the same species whether you are in Colorado, Colombia, or Croatia. In many countries, there is a further distinction made within *C. sativa*, where "hemp" and "marijuana" are legal and cultural designations, but not distinct species, and are regulated differently based on chemical composition.

The taxonomic classification of *Cannabis* has long been a topic of debate. Although *C. sativa* is generally recognized as a single species, additional names and classifications have been introduced to facilitate communication (Table 1). Currently, however, this system remains disorganized. The industry heavily depends on strain names, which are known to lack consistency and are often difficult to verify. Strains are conceptually equivalent to cultivars, with some subtle differences that will be discussed later. Obtaining reliable genetic identification data is still largely difficult and expensive, while chemotype data can be highly variable and not always dependable.

Scientists rely on accurate classification to establish a consistent research approach that enables comparative studies, reproducibility, and the accumulation of knowledge. Cultivators and breeders require clearly defined varieties to develop, preserve, and protect plant traits critical for cultivation and innovation. Medical and regulatory stakeholders also depend on dependable categorization, often organized around chemotype rather than physical appearance, to inform safety, compliance, and therapeutic use. Consumers, by contrast, navigate the market primarily through strain names, which serve more as cultural shorthand than scientific identifiers, shaping a marketplace that is both complex and deeply rooted in tradition.

In the cannabis industry, technical taxonomy might not matter to most people, but language does. Humans have heavily influenced, spread, and shaped cannabis worldwide. Much of the variation observed today results from plasticity, hybridization, and selective breeding, with most wild-growing populations being feral or exhibiting admixture (genetic mixing) from cultivated varieties [3–5]. The flowers of male *Cannabis* plants are prolific pollen producers, with each tiny flower capable of releasing hundreds of thousands of pollen grains that can travel long distances [3], which further promotes admixture.

Unsurprisingly, even though many subscribe to a single species designation, there are still disagreements among botanists on the number of species that should be designated in the *Cannabis* genus. Some botanical experts recognize several species and/ or subspecies [6], while applied fields like medicine, agriculture, and regulation find chemotype and genotype classifications more useful than species distinctions. Additionally, while the legal hemp and marijuana distinctions have no foundation in botanical taxonomy, these groups are now actually well differentiated at the genetic level reflecting genome-wide divergence rather than single-gene variation [5, 7–10] likely because of human selection for different uses, such as hemp for grain and fiber, and marijuana for flowers that are rich in the cannabinoid tetrahydrocannabinol (THC) (Figure 1). The complexity of *Cannabis* taxonomy reflects the interplay of morphological, chemical, genetic, and legal systems, none of which fully resolve the species question (Table 1, Appendix 1).

Terminology that consumers use may be mismatched or less specific than language used in scientific or cultivation settings. Broad labels such as "Sativa," "Indica," and "Hybrid" (Appendix 1) have become popular as cultural branding for effects, rather than as accurate botanical categories. However, these terms no longer reliably

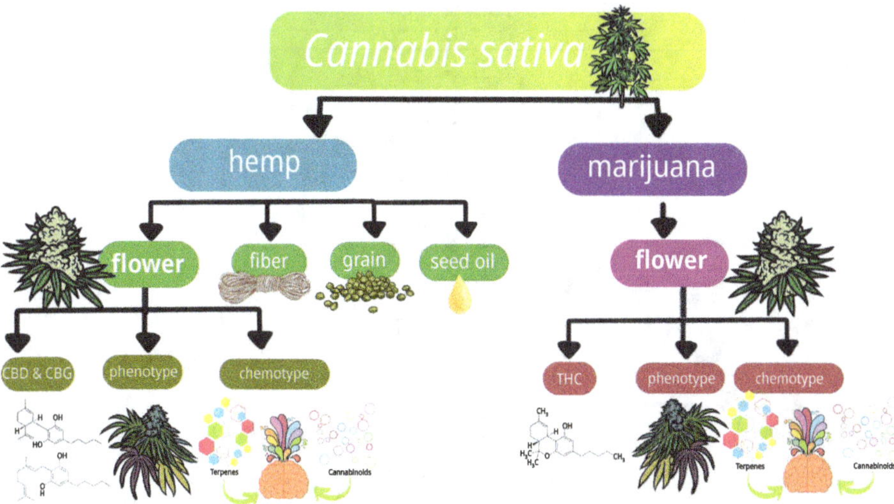

Figure 1: *Cannabis sativa* encompasses both hemp and marijuana, which have been selectively bred for different purposes. Hemp is bred for fiber, grain, seed oil, and low-THC flower rich in cannabinoids such as CBD and CBG, while marijuana is bred primarily for resinous flower high in THC and other psychoactive compounds.

correspond to plant morphology or lineage as they once did, and scientific classification now relies on chemotype, morphology, geography, and ancestry, while growers prioritize growth pattern, chemical profile, and yield (Table 1). Marketing cannabis products as sativa and indica also assumes to predict specific effects, but the reliability is limited. Effects are highly subjective and may be influenced by plant chemistry, environment, and individual physiology [11–14] (Figure 2). Assigning fixed experiences to any single category is misleading, as many variables shape real-world outcomes. The process by which these categories became disconnected from scientific reality is further discussed in the section "Sativa versus indica debate."

This chapter discusses the importance of knowing species, how this relates to *Cannabis*, how we define species, and how categorizing species and categories within species assists with multiple areas in the cannabis industry from research and agricultural practices to consumer understanding and policy decisions. By examining how terms and labels used for cannabis evolved through history and culture, we can better understand where these narratives serve practical needs and where they obscure biological reality (Appendix 1). Rather than rejecting these categories entirely, we need to place them in context, clarifying when they offer useful communication tools and when they fail to capture the complexity revealed by genetics and chemistry. Only by integrating common categories with rigorous data-driven scientific classification can we create meaningful systems that serve researchers, cultivators, regulators, and consumers alike.

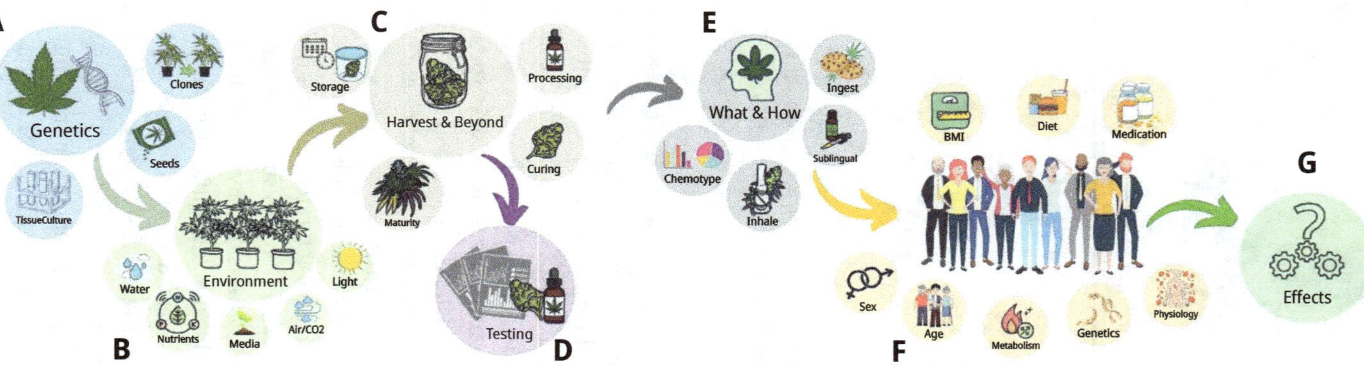

Figure 2: Multiple factors influence the morphology, chemistry, and resulting effects of *Cannabis*. The characteristics and effects of *Cannabis* are shaped by a complex interplay between the plants genetics (A), environment (B), postharvest handling and processing (C), and consumption method (E), all filtered through individual physiology (F). Analytical testing (D) does not directly affect these traits but can influence how they are represented, as differences in sampling, method, or reporting may lead to inconsistent or misleading results. Together, these variables contribute to the wide diversity in plant morphology, chemical composition, and consumer experience (G).

What exactly is a species?

The concept of species is a human construct that helps us communicate about organisms, but there is no universal definition that applies to all life. Nature rarely fits neatly into boxes, and just when we think we understand it, nature surprises us. Multiple concepts have been proposed to define what a species is, and most were developed for and used with sexually reproducing animals [15]. Even among animals, however, boundaries can blur [16]. Some distinct species, such as wolves, coyotes, and domestic dogs, can interbreed and produce fertile offspring [17, 18], while others, like horses and donkeys or lions and tigers, can hybridize but often produce sterile offspring [19]. These examples highlight that even in animals, reproductive isolation is not absolute. Plants often challenge these definitions even further because of asexual reproduction, hybridization, and high phenotypic plasticity. Consequently, botanists often adopt a pluralistic approach that combines genetic, phylogenetic, and ecological criteria to delineate plant species. This approach is especially important for complex groups like those in the *Cannabis* genus, which readily hybridize, exhibit phenotypic plasticity, and have been greatly influenced by selection and breeding.

One of the most widely used models is the **Biological Species Concept**, first proposed by Ernst Mayr in 1942, which defines a species as a group of actually or potentially interbreeding populations that are reproductively isolated from others [20]. While this concept works well for animals, it often falls short in plants due to frequent hybridization in closely related taxa. The **Morphological Species Concept**, introduced by Arthur Cronquist in 1978 classifies species based on consistent and distinctive physical traits, such as leaf shape, flower morphology, stature, or branching patterns [15]. While this approach helps identify fossils and specimens in the field, it may be less reliable for organisms whose characteristics change with environmental or cultivation conditions. This reliance on physical traits underpinned the early taxonomic split between *Cannabis* species, as discussed in the following section. The **Ecological Species Concept** defines species by their adaptation to specific ecological roles [21], while the **Phylogenetic Species Concept** identifies species as the smallest diagnosable groups on an evolutionary tree, emphasizing shared ancestry and descent [22, 23].

Applying these concepts to *Cannabis* shows the complexities involved. Historically, botanists identified separate species *C. sativa* L. [24] and *Cannabis indica* Lam. [25] based on differences in morphology and geography. Later, Small and Cronquist [26] merged them into one species, *C. sativa* L., with subspecies or varieties such as *C. sativa* subsp. *sativa* (fiber and seed types) and *C. sativa* subsp. *indica* (drug types). Since then, botanists have debated these classification suggestions. Some argue that maintaining a taxonomic split could even have legal implications, since early bans in the United States only mentioned *C. sativa*, opening the argument that other types, like *C. indica* or *Cannabis ruderalis,* were not officially prohibited [4]. This argument does not appear to have been successfully applied in court (e.g., United States v. Rothberg [27] and People

v. Van Alstyne [28]), but it illustrates how taxonomy and law became intertwined in *Cannabis* history. The claim that the Controlled Substances Act (CSA) only prohibits *C. sativa* has been raised several times but to my knowledge has never influenced a legal outcome. Courts consistently rule that the legislative intent was to outlaw all intoxicating forms of *Cannabis*, not a single species. These rulings show that early courts interpreted *"Cannabis sativa L."* in the CSA and earlier statutes such as the Marihuana Tax Act (1937) as reflecting the scientific consensus of the time, which recognized only one species. As a result, taxonomic distinctions have not been effective in influencing enforcement or legal definitions.

Modern genetic research supports the idea that *Cannabis* is best treated as a single, highly diverse species with structured populations shaped by domestication, hybridization, and human selection rather than showing clear reproductive boundaries between different species. However, there may be a compelling argument for further division of the species into subspecies or varieties as several taxonomic treatments have proposed [6]. Researchers have suggested that the morphological, geographic, and chemical diversity within *Cannabis* may warrant recognition of infraspecific taxa such as *C. sativa* subsp. *indica*, *C. sativa* var. *spontanea*, and other domesticated or feral forms [4, 26, 29, 30]. While modern genetic data often support a single, highly variable species, these historical frameworks provide a useful foundation for understanding the range of diversity shaped by domestication, selection, and regional adaptation.

Establishing formally recognized nomenclature of plants

The pathway to naming plant species follows an internationally recognized set of rules established by the ICN for algae, fungi, and plants . This system guarantees that each species has a unique, validly published scientific name and that these names are used consistently throughout the scientific community [31]. The ICN is governed by the International Botanical Congress, which convenes every six years to review and update the code.

For a name to be officially recognized, taxonomists must publish a valid description in a peer-reviewed journal, designate a type specimen, and deposit that specimen in a recognized herbarium. The type specimen serves as the permanent reference point for that species name, enabling future researchers to compare new material to the original description. This process ensures that names are not merely cultural conventions but are connected to physical evidence and community consensus. It is a rigorous system designed to maintain consistency in naming and establish universal standards for communication. In contrast, names used in cannabis breeding and commerce have mostly developed outside this system. Strain names are not linked to type specimens, and there is no requirement for valid publication or formal review. As a result, *Cannabis* has accumulated thousands of names [32–35] that hold cultural significance but lack the formal stability and reproducibility that botanical nomenclature provides. No comprehensive or standardized database exists. However, commercial and crowdsourced resources such as SeedFinder and industry directories such as Leafly, Weedmaps, and Canna Connection list tens of thousands of entries [32–35]. Even genetic repositories and research datasets contain hundreds of unique submissions from seedbanks, breeders, clone suppliers, and dispensaries, underscoring the fragmented nature of *Cannabis* nomenclature.

In addition to the naming of wild plant species under the ICN, cultivated plants are governed by a separate system: the International Code of Nomenclature for Cultivated Plants (ICNCP) [1]. This code sets rules for naming cultivated varieties (aka cultivars) that have been intentionally selected and maintained by humans. To be officially recognized, a cultivar must show distinct, uniform, and stable traits. The name must follow ICNCP conventions, which typically include a Latin binomial for the species followed by a cultivar epithet in single quotation marks (e.g., *Vitis vinifera* 'Cabernet Sauvignon' or *Malus domestica* 'Granny Smith'). Unlike species names, cultivar names are not italicized. Cultivar registration is managed through International Cultivar Registration Authorities (ICRAs), which are assigned to specific plant groups. These authorities keep official records, prevent duplicate names, and ensure that recognized cultivars can be traced, preserved, and reliably propagated [36].

Cannabis has not historically been part of this system because prohibition kept the plant out of formal agriculture, and breeders lacked access to ICRAs or cultivar registration pathways. Instead, names like Skunk, Haze, or OG Kush spread informally through underground markets without formal oversight or type material. This led to the current landscape where thousands of strain names exist without standardized definitions or verification. However, as legal markets grow, there are increasing efforts to align *Cannabis* cultivars with ICNCP principles, especially for breeder rights, intellectual property (IP), and germplasm preservation.

Taxonomic treatment of *Cannabis*

In taxonomy, species is the most specific rank, generally defined as a group of organisms capable of interbreeding. While many species maintain reproductive isolation, plants often cross these boundaries. In *Cannabis*, as in many plants, interbreeding between populations and purported species and subspecies is common, which supports treating the genus as a single, highly diverse species rather than multiple distinct ones.

Early taxonomists classified *Cannabis* as multiple species based on morphology and geography. Linnaeus (1753) described *C. sativa* as tall plants with sparse branching, loose flower structures, and narrow leaflets [24]. Lamarck (1785) later described *C. indica* as shorter plants with dense branching, compact flowers, and broader, darker leaves [25]. In 1924, Janischewsky proposed *C. ruderalis* for small, weedy, early flowering populations from northern and central Eurasia adapted to short growing seasons [37] (Figure 3).

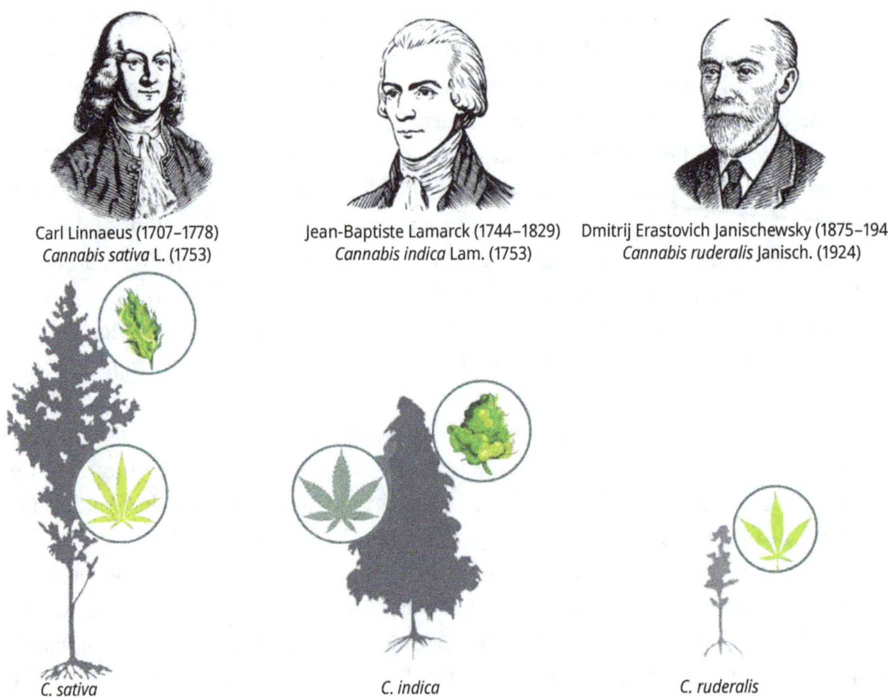

Figure 3: Historical taxonomic classifications of *Cannabis*. Carl Linnaeus (1753) first described *Cannabis sativa* based on tall, narrow-leaf plants cultivated in Europe. Jean-Baptiste Lamarck (1785) later identified *Cannabis indica* from shorter, broad-leaf, resinous plants from India. Dmitrij Erastovich Janischewsky (1924) subsequently described *Cannabis ruderalis* from weedy, early-flowering populations in central Russia. These three descriptions form the foundation of modern debates over the taxonomy of *Cannabis sativa*.

Historically, *C. sativa* types were associated with warmer lowland regions and cultivated mainly for fiber and seed. In contrast, *C. indica* types were linked to colder, higher-altitude areas where compact, resin-rich plants were used for drug production. Classic examples include tall, airy "sativa-like" landraces from Thailand and other equatorial regions, and compact, resinous "indica-like" landraces from Afghanistan and the Hindu Kush Mountain range. Geography and climate clearly influenced these phenotypes, but the idea that *C. sativa* equals "hemp" and *C. indica* equals "drug" has been more of a taxonomic assumption or usage application rather than a strict biological fact.

The modern commercialization of *Cannabis* has increasingly blurred any distinctions that may have previously existed. Decades of intensive breeding, widespread hybridization, and the incorporation of *C. ruderalis* genes for auto-flowering traits have further complicated the picture and much of the historical differentiation that was once evident in morphology has been muddled [38]. Genetic studies using molecular markers, single-nucleotide polymorphisms, and sequencing techniques reinforce this overlap [5, 8–10, 38–40]. Importantly, the common categories of Sativa and Indica do not align with consistent genetic clusters. Evidence also suggests that domestication occurred multiple times across Eurasia [41], resulting in both fiber- and drug-type biotypes, rather than clear species divisions.

Cannabis can also be classified based on their chemical makeup and/or intended use [3, 42–45]. Major cannabinoids like THC, cannabidiol (CBD), and cannabigerol (CBG) can be used to define five main chemotypes of cannabis intended for consumption. These categories separate plants into THC-dominant (type I), balanced THC:CBD ratio (Type II), CBD-dominant (Type III), CBG-dominant (type IV), and cannabinoid-null (type V) [46–48]. While useful in medicine and regulation, these groupings simplify diversity to just a few key cannabinoids. *Cannabis* varieties that have been bred for secondary cannabinoids such as cannabichromene (CBC), tetrahydrocannabivarin (THCV), and the newly discovered, highly potent tetrahydrocannabiphorol (THCP), along with terpenes, flavorants, esters, aldehydes, flavonoids, and thiols are gaining interest, and these five biotypes are simplistic and now not as useful as when they were proposed. These other compounds also influence aroma, flavor, therapeutic effects, and user experience [49–52]. Therefore, while these five proposed chemotypes offer a practical guide for regulatory and clinical use, they do not represent the full biochemical complexity nor consider the breeding histories that shape modern cultivar diversity.

Ultimately, no single system fully resolves *Cannabis* taxonomy. Morphology, geography, and chemistry intersect in complex ways, and each only explains part of the story. Table 1 combines these perspectives, showing how the traditional taxonomic categories overlap with chemotypes and pointing out the limitations of each approach.

Historical context of modern *Cannabis* taxonomy

Twentieth-century prohibition in the United States and worldwide pushed *Cannabis* into informal and underground markets, influencing how plants were named, classified, and traded. Without access to germplasm repositories, cultivar registration, or breeder IP protections that are available for other crops, *Cannabis* clones and seeds were passed informally, with strain identities emerging through word of mouth, underground magazines, and early seed catalogs, rather than standardized descriptors or verified pedigrees.

Before prohibition tightened globally, locally adapted landraces, such as Thai, Afghan, Hindu Kush, Lebanese, and Moroccan, were transported as seeds by travelers and early collectors, forming the first Western breeding stock. With no official seed banks available, private collectors and underground seed companies became the primary sources of genetic material during this formative period [30]. Legend has it that two figures were particularly influential during this era: David Watson ("Sam the Skunkman") and Nevil Schoenmakers ("The King of Cannabis"). Watson is credited with developing legendary cultivars including Skunk #1 and Original Haze before relocating to Amsterdam in the mid-1980s. Schoenmakers founded The Seed Bank of Holland in 1984, which reportedly was the first cannabis seed company to openly distribute seeds worldwide through mail-order catalogs. The collaboration between these early breeders helped establish foundational genetics that would be repeatedly recombined throughout the industry. Due to limited legal access to fresh germplasm, breeders repeatedly recombined the same genetic families. This process led to the first significant wave of polyhybrids, which were created by crossing multiple hybrid lines with diverse traits inherited from the parents. This relatively narrow genetic base established by early pioneers (Figure 4) is thought to be the genetic basis for the vast majority of commercial cannabis varieties available today [4, 7].

Underground cultivation, breeding, and trading took place without formal oversight, and documentation standards varied greatly. Exchange depended on informal networks, gray-market seed sales, and clone swapping. Selection was based on survival and demand, which included potency (not THC content), resin production, flowering cycle, yield, and aroma rather than taxonomic stability. Phenotype hunting or "phenohunting" of segregating seed lots produced "keepers," which were then shared as named clones without genetic verification. With no practical path to cultivar registration or breeder IP protection because of the illegal status of the plants, there was little incentive to develop homozygous parental lines or stable F1 hybrids like other crops. Instead, clonal propagation became standard for maintaining elite genotypes. Today, genetic tools for verification and tissue culture have greatly improved the potential to maintain and track genetic lineages, as well as enabling development of disease-indexed clones, and long-term preservation of valuable genotypes but is still outside the structure of formal cultivar registration.

The legacy of this history is a market with tens of thousands of named strains, many derived from a narrow gene pool and shaped by repeated hybridization. Seedfinder is by far the largest and most comprehensive to date, with over 37,000 different named cannabis varieties [32]. Without formal naming or verification systems, folk nomenclature filled the gap. Terms such as "Sativa," "Indica," "OG," countless strain names, and vague quality indicators, such as "kind," "dank," "schwag," "brick," or "ditch weed," became the working language of the underground market. While this vernacular enabled communication among growers and consumers during prohibition, it also established the foundation for current issues with mislabeling, duplication, and inconsistent product identity. The modern marketplace remains rich in names and cultural terminology but lacks standardized, testable reference points, which is a problem explored further in the section "Strain names and market confusion."

This legacy explains why modern commercial labels communicate culture and expectation more than consistent biology. Prohibition entrenched a system where taxonomy was improvised, breeders worked within narrow constraints, and names spread without verification. By contrast, many other crops heavily shaped by human selection progressed through formal agricultural and taxonomic systems, leading to stability and clarity. A frequent debate in cannabis underscores this point: advocates argue for "cannabis" rather than "marijuana," yet hemp is also *Cannabis*. To illustrate this, it would sound odd to say, "We are having brassica for dinner tonight." The name of the genus alone carries little meaning in daily life. Instead, we distinguish among the many botanical varieties of *Brassica oleracea* including broccoli, cauliflower, cabbage, kale, and Brussels sprouts. None of these vegetables exist in the wild in their familiar forms. There is no such thing as "wild broccoli" (Figure 4). Because their development took place within legal agricultural systems, these crops were stabilized, registered, and recognized as varieties of *B. oleracea*. By contrast, *Cannabis* has thousands of unverified strain names which may be culturally meaningful but are often inconsistent in lineage and/or chemistry and clearly highlights why formal taxonomic systems matter.

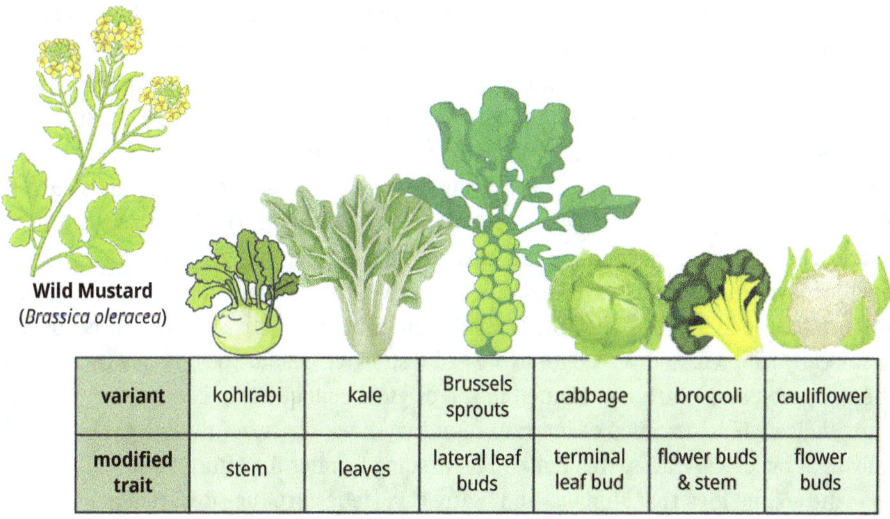

variant	kohlrabi	kale	Brussels sprouts	cabbage	broccoli	cauliflower
modified trait	stem	leaves	lateral leaf buds	terminal leaf bud	flower buds & stem	flower buds

Figure 4: Selective breeding of *Brassica oleracea* (wild mustard) has produced multiple cultivated forms, each emphasizing a different plant part. Kohlrabi was bred for its enlarged stem, kale for its leaves, Brussels sprouts for lateral leaf buds, cabbage for the terminal leaf bud, broccoli for flower buds and stem, and cauliflower for compact flower buds.

The "Sativa" versus "Indica" debate

Few topics in *Cannabis* spark more debate than the mention of "Sativa" and "Indica" (Appendix 1). A phrase famously coined by Kristen Yoder "The Cynical Stoner," "Sativa/

Indica is lazy marketing." However, consumer culture has perpetuated these terms and become convenient labels for expected effects rather than accurate reflections of plant biology. People often share the concept that a "Sativa" strain is uplifting, invigorating, and energizing, while an "Indica" strain is sedating, relaxing, and calming. Although rooted in previous taxonomic nomenclature, their meanings have shifted with consumer use and marketing, drifting away from their scientific definitions. Today, most consumers use these terms based on experience, not observable plant traits.

I argue though, that although reductive and somewhat arbitrary, these terms are not meaningless as many folks forcefully declare. They are deeply entrenched in cannabis vernacular and offer a shared language for consumers, budtenders, and growers. While once rooted in science and plant structure, their meanings have evolved. Today, many assume that "Sativa" and "Indica" reliably signal distinct effects, but available evidence suggests these categories are not consistently supported by chemistry or genetics [45]. Clear genetic clusters for these common groups have not been demonstrated, and chemical analyses often show substantial overlap in cannabinoid and terpene profiles across categories. Smith et al. [45] did find that products labeled "Sativa" were somewhat more likely to fall into a terpene cluster characterized by high terpinolene and myrcene, but this association was modest and not observed for other terpene groupings. Despite these scientific limitations, the terms persist as familiar communication tools in commercial and cultivation contexts.

It is also important to recognize that current conclusions are drawn largely from cannabinoids and terpenes. As previously mentioned, other classes of compounds such as flavonoids, esters, thiols, and minor cannabinoids like CBC or THCV may contribute to perceived differences and remain underexplored. Perceived differences themselves add another layer of complexity, as they are subjective, context-dependent, and influenced by variables such as sleep, diet, medications, and setting [11–14]. Moreover, it is possible that locally adapted landraces once expressed distinctive chemotypes shaped by environmental pressures, which contributed to the original colloquial divide between "Sativa" and "Indica." If so, decades of hybridization may have obscured those signals, leaving only faint biological traces. This remains an exciting frontier for future research.

And so, the idea that "Indica" and "Sativa" varieties produce different experiences might not be entirely unfounded. Many people say they can distinguish between them, even if the differences are more nuanced than just "sedating" versus "uplifting" dichotomy. There could be unidentified genetic markers or chemical compounds influencing these perceptions. A parallel can be drawn to the golden-winged warbler (*Vermivora chrysoptera*) and the blue-winged warbler (*Vermivora cyanoptera*). Despite being almost identical in their genomes (99.97% identical, with only six regions showing significant differences), small genetic variations explain the visible traits and behaviors that justify classifying them as separate species [53]. Similarly, in *C. sativa* may share of the genome across varieties, but minor genetic differences or the presence of trace yet potent compounds could influence the effects people associate with "Indica" and "Sativa".

It is also reasonable to theorize that regional environments may also have influenced the evolution of *Cannabis* phytochemical production (Figure 5). *Cannabis* exposed to different pests, diseases, or temperature stresses can develop unique chemical defenses in response to local pressures. For example, plants growing in colder or high-altitude climates, or under intense sunlight, may produce higher levels of flavonoids like anthocyanins, which protect them from freezing damage and excessive UV light exposure (Figure 5). Many of these compounds are still largely unexplored, but they might play a role in the unique experiences consumers report from various regional types.

The bottom line is that the concepts of "Sativa," "Indica," "Hybrids," and strain names are cultural tools for understanding a complex plant. They are not strict scientific categories, but they serve as a common language. The goal now is not to dismiss or eliminate them but rather to put them into context, matching cultural shorthand with the expanding body of genetic and chemical data, so that communication remains both accessible and precise.

Figure 5: Potential regional adaptation and chemical diversity in *Cannabis sativa*. *Cannabis* grown in different regions may face varying environmental pressures, including pests, pathogens, temperature, and sunlight, resulting in small adaptive genetic changes and the development of chemical defenses. These subtle differences could help explain why people still perceive distinct effects between varieties.

Strain names and market confusion

If "Sativa" and "Indica" are the cultural shorthand for effects, then strain names serve as the everyday touchstones of *Cannabis* identity. These names emerged through folk taxonomy, underground marketing, and regional traditions rather than formal standards. Vintage strains like Panama Red, Mazar, Durban Poison, Maui Wowie, Chemdawg, Jack Herer, and Purple Kush conveyed regional roots, cultural significance, and sensory expectations, often linked to the reputation of a breeder or area. Branding and storytelling added layers, making names a key part of the consumer experience, just as important as the product itself. Modern strain names like Girl Scout Cookies

(GSC), Wedding Cake, Zkittlez, Blue Dream, Runtz, Gorilla Glue #4 (GG4), Lemon Skunk, and Blueberry Cookies (Blueberry × GSC) do more than identify a product, they signal expected flavors, effects, and sometimes even hint at the breeder's reputation or specific genetic background to consumers. These names reflect how the modern *Cannabis* markets combine marketing, cultivation history, and consumer experience, turning each strain into a story that merges unique sensory qualities, breeder identity, and lineage into a single label. However, the lack of official registration systems and IP protections for *Cannabis* strains and cultivars, which are standard for breeding and marketing in other crops, leads to frequent renaming, mislabeling, and uncertainty about product identity. Therefore, the strain name on a package does not always match what is inside [38]. Moreover, names are important because they facilitate communication and establish expectations. For example, a name like Lemon Haze indicates that the flower may have a citrus aroma and an uplifting effect. Yet, names can shift, drift, be duplicated, or be misused, which has also been a common issue in other agricultural commodities until genotyping tools existed and/or registration systems were established [54–58].

In traditional agriculture, breeders who develop a new cultivar, such as the tomato variety 'Better Boy,' can officially register it under plant variety protection laws and earn royalties or licensing fees from seed companies that sell the seed through the United States Department of Agriculture (USDA) Plant Variety Protection Office. This legal system ensures proper naming, genetic verification, and compensation for breeders, while providing buyers with confidence that the product matches its label. Since these systems generally have not existed for *Cannabis*, breeders are not afforded the same pathway to protect unique varieties. Names like OG Kush or GSC can be used freely and sometimes incorrectly, without oversight. This has resulted in inconsistent genetics and confusion among consumers.

It would be remiss to discuss the debates over language to not also include the word "strain." Some people insist that strains are limited to microbes, and that "cultivar" is the correct botanical term for plants. A cultivar, short for "cultivated variety," refers to a plant variety produced through selective breeding for specific traits. Cultivars, by definition, have stable and genetically verified lineages and can be reliably reproduced from seeds or clones within formal agricultural systems [1]. Grapes, apples, and olives, for example, all have cultivar names tied to verified lineages, so consumers can trust that 'Granny Smith' or 'Cabernet Sauvignon' always refer to the same variety. *Cannabis*, however, often does not meet these requirements, especially when grown from seed, largely due to polyhybrid ancestries.

In cannabis, the word "strain" is used inconsistently. It might refer to a pack of seeds from a cross, where each seed represents a unique genotype and can express a different phenotype. It might instead refer to a single mother plant and all of the genetically identical clones propagated from it (and subsequent generations of those clones). In other cases, it may even describe seed derived from sibling crosses or from self-fertilization using techniques such as chemical reversal ("feminization"). In practice, a "strain" can also represent an entire lineage. Permanent Marker, for example, was a 2022 release from Seed Junky Genetics and Doja Exclusives, and is reportedly a three-way cross of Biscotti S1, Sherb BX, and Jealousy. Yet in practice, the name now refers to a family of related cuts and seed lines derived from that original breeding. In this sense, "Permanent Marker" is not a single uniform genetic entity, but a branded lineage whose identity is maintained more by cultural recognition and market adoption than by genetic or chemical consistency [32, 59].

Therefore, a "strain" may have limited information about its genetic stability or how it should be propagated to maintain the desired traits over generations. Moreover, it is important to recognize that "strain" is the most used term in the cannabis community to describe groups of plants with a shared lineage. Unlike registered cultivars, however, most strains are polyhybrids that are not consistently maintained or genetically stable. Their primary function is to signal lineage and identity within the community, even if they lack the uniformity expected of formal cultivars. The ICNCP proposes "cultigen" as an alternative term, referring to a systematic group of cultivated plants [60]. While accurate, it has not gained the widespread popularity or everyday usage that strain has within the cannabis community, and strain remains the dominant term in the industry...an imprecise label, but not entirely incorrect.

Grapes, apples, and olives all have cultivar names tied to verified lineages, so consumers can trust that 'Granny Smith' or 'Cabernet Sauvignon' always refer to the same variety. *Cannabis* strains lack this standard. Different plants may be sold under the same name, sibling plants from the same seed lot may all carry that name despite differences, and genetically identical clones are often circulated under multiple names.

Scientific studies have verified this disconnect. Genetic analyses, including my own research, show that samples sold under a specific strain name often do not cluster genetically, while in other cases, genetically identical plants are sold under completely different names. Chemical profiles can also vary greatly within a named strain, influenced by growing conditions, harvest timing, and post-harvest handling. This variability hampers reproducibility. For recreational users, the effects may be limited to inconsistent experiences, but for medical patients seeking dependable symptom relief, this inconsistency is a significant obstacle to market maturity.

The absence of standardized naming also makes it harder to protect IP and gain breeder recognition. Without consistent definitions or official registration systems, breeders have limited ability to safeguard their work, and consumers lack confidence that what they purchase matches the name on the label. Cloning has offered one solution for preserving elite phenotypes, and tissue culture now provides a more advanced option for maintaining disease-free lines. However, even these methods operate outside a formal system of cultivar registration or genetic verification.

Currently, we have a market full of names but lacking consistent genetic validation. Strain names remain crucial as cultural indicators, helping consumers and communities share preferences, but they cannot ensure biological or chemical uniformity. The answer is not to discard these names but to enhance them with systems that connect names to verifiable data. Techniques like genotyping, chemoprofiling, and breeder verification platforms are emerging to offer that scaffold. Over time, this alignment between names and scientific validation could make *Cannabis* more consistent, combining tradition and identity with reproducibility and trust.

Genetic realities: what we actually know

Modern genomic research has provided a clearer, though more complex, picture of *Cannabis* diversity. Broadly, *Cannabis* typically divides into two main genetic clusters that roughly correspond to hemp and marijuana types. While some argue these groups could be considered subspecies, most genetic studies agree on one key point: the boundaries between *C. sativa* and *C. indica* are not sharply defined, nor are the colloquial "Sativa" and "Indica" labels. Instead, gene flow and extensive human-driven breeding have created a genetic continuum rather than clear-cut categories.

Adding to this complexity is the incorporation of *C. ruderalis* genetics. Originally described as small, weedy populations from northern Eurasia, *C. ruderalis* contributed early-flowering and auto-flowering traits that have become valuable in modern hybrids. These traits are especially useful in short-season climates and indoor cultivation, where faster cycles and reduced space requirements are advantages. The introduction of these genetic traits is likely further blurring any remaining distinctions among the groups based on morphology or geography. Because *C. ruderalis* is genetically closer to hemp, its widespread use in breeding may also erode the already thin line between fiber- and drug-type populations.

Within this continuum, chemotypic groups (Types I–V) become more distinct than traditional species classifications. These categories have proven valuable in medicine and regulation because they align with pharmacological and legal outcomes. However, while chemotypes are helpful, they are also limited because they only represent a small portion of *Cannabis* chemodiversity. Besides THC and CBD, cultivars can produce other cannabinoids and compounds that influence aroma, flavor, and potentially effects. Chemotypes indicate which cannabinoids are dominant, but they do not fully capture the biochemical or experiential diversity seen in real-world *Cannabis* plants and experiential influences.

Genomic research also emphasizes the importance of differentiating between genotype and phenotype. A single genotype can produce varying chemical results depending on environmental factors such as light spectrum, soil nutrients, pest pressure, and post-harvest handling. Epigenetic regulation introduces another layer, as gene expression can change without altering the DNA sequence itself. This explains why genetically identical clones can show noticeable differences in terpene or minor cannabinoid content when grown in different environments. The resulting phenotypic plasticity has direct consequences for both recreational users and medical patients. Inconsistent chemical profiles can cause unpredictable effects, making it hard for consumers to replicate experiences from one purchase to another. For medical users, this inconsistency may affect symptom management and reduce confidence in cannabis products as reliable therapeutic options. Ultimately, environmental and epigenetic factors pose ongoing challenges to achieving accurate labeling and consistent outcomes in both the hemp and high THC sectors of the cannabis industry.

Overall, the genetic evidence not only indicates unity within the genus but also high variability. *Cannabis* is best understood not as multiple species but as one species with structured populations, overlapping chemotypes, and an extraordinary capacity for hybridization. This complexity challenges simple categorizations and naming schemes but also creates opportunities. By linking genotype, chemotype, and environment, researchers and breeders can begin to map the relationships that matter most for cultivation, medicine, and consumer experience.

Why classification matters and how to proceed

Although *Cannabis* does not resolve neatly into distinct species, classification is still important. Differentiating species and varieties forms the basis for how humans understand, organize, and communicate about plants. In agriculture and horticulture, species labels are essential for comparing research, developing breeding programs, and establishing regulatory systems. Without a shared classification system, results cannot be reliably reproduced, cultivars cannot be protected, and policies cannot be effectively enforced.

The need for reliable *Cannabis* classification has wide-reaching effects along the entire supply chain in the cannabis industry. For cultivators, accurate variety identification guides choices in crop management, harvest timing, and disease prevention. Breeders need stable names and identities to ensure the legacy and progress of their work. For medical patients and practitioners, consistent labeling is essential to achieving safe and effective therapeutic outcomes. Regulators rely on clear definitions to create fair laws and protect consumer safety. While federal descheduling of all plants within *C. sativa* could remove specific regulatory differences between industrial use *Cannabis* (hemp) and consumable *Cannabis* (regardless of THC content), it would not resolve challenges faced by cultivators, breeders, or consumers. Accurate differentiation and classification remain vital for maintaining trust, supporting innovation, and delivering consistent experiences to patients and consumers.

At the same time, misunderstandings exist about what classification can and cannot do. *Cannabis* hybrids and polyhybrids are common, and labels alone cannot represent the full diversity in the current market gene pool. Moving forward, I believe *Cannabis* reconciliation could benefit from a dual-classification system. On the one hand, vernacular categories like "Sativa," "Indica," and familiar strain names serve as useful communication tools, allowing consumers, budtenders, and cultivators to speak a common language. On the other hand, science and policy require data-driven evidence rooted in chemotypes and genotypes. Classifying chemovars (chemical variants) based on cannabinoid and terpene profiles provides potential practical groupings for medicine and regulation but may add consumer confusion if they are required to learn the vast array of chemical compounds in *Cannabis* that influence effects. Most consumers don't want or care to know every ingredient that goes into a dish, and similarly, they don't care about the specifics of the chemical profile of the flower they enjoy. Genomic tools and databases can establish verified identities for cultivars, ensuring consistency across markets, and connecting chemistry, genetics, and sensory experience using real world data will create a multidimensional classification that reflects identity and complexity.

The future of *Cannabis* should not involve eliminating cultural language but rather placing it within a stronger scientific organization using evidence-driven data. A hybrid approach that respects existing vernacular while incorporating genotype- and chemovar-based systems can clarify an industry still heavily influenced by the legacy of prohibition and decades of informal naming. By aligning communication with science, the cannabis industry can become more transparent, reproducible, and trustworthy, benefiting consumers, patients, breeders, and regulators alike.

Conclusion

Cannabis taxonomy has always been complex. Historical classifications based on morphology divided the plant into multiple species, while modern genetics indicate that it is best understood as a diverse species shaped by domestication, hybridization, and culture. The continued use of "Sativa," "Indica," and strain names reflects how communities improvised classification systems under prohibition, creating a common vernacular that still dominates consumer markets. Meanwhile, science has shown that these categories do not necessarily align with genetics or chemistry, and their usefulness is limited to communication rather than prediction.

Knowing plant species matters because classification influences how we share knowledge, regulate markets, breed and protect cultivars, and provide consistent experiences for patients and consumers. And although for *Cannabis*, we are only looking at a single species, the current state of *Cannabis* nomenclature debates demonstrates what happens when classification is driven by culture and underground markets, rather than being incorporated into formal agricultural and scientific systems. The abundance of named strains and clear taxonomy is not necessarily a bad thing; we just need to approach this with understanding. Legacy breeders and growers built this system without regulation and scientific rigor, but we also need to recognize that science professionals only recently took a seat at the table. To erase the history of the illicit industry fueled by prohibition and the War on Drugs is, in my view, disrespectful to those whose shoulders we stand on in building this billion-dollar industry.

Systems applied to other crops demonstrate what is possible when domestication occurs within legal and scientific frameworks. As noted earlier, none of the familiar forms of *B. oleracea* sold in supermarkets exist in the wild, yet because they have been stabilized, registered, and preserved through formal agricultural systems, they remain consistent and recognizable worldwide. Consumers can trust what they are purchasing, breeders can protect and improve germplasm, and researchers can compare results across time and regions. By contrast, the *C. sativa* forms, we are familiar with were developed under prohibition without such a system, producing a patchwork of names and identities rather than a coherent taxonomy.

Although classifications of *Cannabis* species, subspecies, varieties, cultivars, and strains may not fit perfectly into strict categories, it is possible to achieve the stability and reproducibility seen in other crops when vernacular language is combined with science-based classification. The integration of cultural categories with genomic and chemical data is not only feasible but also essential. This approach will help understand *C. sativa* as a single species with remarkable internal diversity, organized in ways that support science, medicine, cultivation, and culture.

Table 1: Taxonomic and chemotypic perspectives on *Cannabis.*

Category/ type	Morphology and ecology	Geography/ adaptation	Traditional usage	Chemistry (major and secondary compounds)	Modern limitations
C. sativa (Linnaeus, 1753)	Tall, narrow-leaf, sparse branching, loose flowers	Warm lowland/ equatorial regions (Europe, South-East Asia, and Africa)	Fiber, seed, and some drug types	Often CBD-dominant (type III) but also THC-dominant (type I); can produce CBC, CBDV	Morphology not predictive of chemotype; "sativa = uplifting/hemp" is cultural shorthand
C. indica (Lamarck, 1785)	Short, broad-leaf, dense branching, compact flowers	Colder, high-altitude regions (Hindu Kush and Indian subcontinent)	Resin-rich drug production	Mostly THC-dominant (type I) but can be balanced (type II); THCV enriched in some landraces	"Indica = sedating" oversimplifies; hybrids blur distinctions
C. ruderalis (Janischewsky, 1924)	Small, weedy, early-flowering, few branches	Central and northern Eurasia; short-season adaptation	Minimal traditional use and source of auto-flowering	Sometimes CBG-dominant (type IV) or cannabinoid-null (type V)	Often treated as biotype rather than species; valued mainly in breeding

Table 1 (continued)

Category/ type	Morphology and ecology	Geography/ adaptation	Traditional usage	Chemistry (major and secondary compounds)	Modern limitations
Modern hybrids	Wide range of morphologies; polyhybrid dominance	Global distribution; shaped by intensive breeding	Recreational, medical, and industrial	Encompass all chemotypes (I–V) plus secondary cannabinoids (CBC, CBDV, THCV, and THCP) and diverse terpenes, esters, aldehydes, flavonoids, and thiols	Neither morphology nor chemotype alone explains diversity; consumer terms persist
Chemotype (Types I–V)	Not morphology-based	Not geography-based	Medical, regulatory, and industrial categories	Type I: THC-dominant; type II: balanced THC:CBD; type III: CBD-dominant; type IV: CBG-dominant; type V: cannabinoid-null	Useful shorthand for medicine/ regulation but reductive; ignores lineage, environment, and secondary metabolites
Major terpenes	Not morphology-based	Not geography-based	Medical, regulatory, and industrial categories	Most abundant (usually three) mono- and sesquiterpenes dominating the profile	Useful shorthand for medicine or organoleptic preferences but reductive; ignores lineage, environment, and secondary metabolites

Appendix 1

Cannabinoid: Any of a diverse class of chemical compounds produced by cannabis that interact with the body's endocannabinoid system (e.g., THC, CBD, and CBG).

Cannabis, cannabis: Colloquial use generally used when talking about *C. sativa* flower and products that are rich in THC. Used like "dog" is used for the species *Canis familiaris.*

Cannabis: The *Cannabis* genus.

Cannabis indica **Lam.**, *C. indica:* The formal scientific name for the species described by Lamarck, often used to describe compact, resin-rich plants purportedly adapted to higher elevations or varied climates.

Cannabis industry: The commercial system encompassing the cultivation, processing, regulation, and sale of cannabis products for medical, recreational, and industrial purposes.

Cannabis ruderalis **Janisch.**, *C. ruderalis:* The formal scientific name for the species described by Janischewsky, typically characterized by small, weedy, early-flowering plants adapted to short growing seasons in central and northern Eurasia. These plants flower based on age rather than the light cycle, and they have played a crucial role in breeding modern auto flowering strains. Its taxonomic status remains debated, and it is often treated as a biotype or ecotype rather than a true species.

Cannabis sativa **L.**, *C. sativa:* The formal scientific name for the species described by Linnaeus, including most cultivated fiber, seed, and drug-type forms.

Chemodiversity: the diversity of chemical compounds produced and in *C. sativa*, refers to the variation in cannabinoids, terpenes, and other metabolites that arise from genetic and environmental differences.

Chemotype: A chemically distinct entity within a species, defined by its unique chemical profile, especially dominant cannabinoids and terpenes.

Chemovar: Short for "chemical variety," it is a chemically distinct entity within a plant species, distinguished by differences in the composition of secondary metabolites (Clarke).

Cultivar: Short for "cultivated variety." Refers to a plant variety bred and maintained by humans for specific characteristics and capable of stable propagation through seeds or cloning.

Genotype: The genetic makeup of an organism; the heritable blueprint that together with the environment determines the phenotype.

Hemp: A legal term for *C. sativa* bred for and grown primarily for fiber, seed, or non-intoxicating cannabinoids (like CBD and CBG), often defined by very low THC content (< 0.3% in the U.S).

Hemp industry: The sector devoted to the cultivation, processing, and commercialization of hemp for fiber, food, and cannabinoid extracts.

Hybrid: results from the cross of two genetically distinct parents producing offspring that express a mix of characteristics from both lineages. However, the term is also widely used colloquially to describe products or effects (see **Sativa, Indica, Hybrid**

below), and this dual usage (genetic versus experiential) leads to confusion, as "hybrid" effects do not necessarily reflect true genetic ancestry.

Marijuana: A nonscientific, historically political term for cannabis plants or products that contain sufficient THC to produce intoxicating effects. Commonly used in legal and regulatory settings to distinguish high-THC cannabis from hemp, though it has no botanical validity. The term carries cultural and legal connotations and is sometimes avoided in scientific discourse in favor of "cannabis."

Phenotype: The observable physical, biochemical, and behavioral traits of an organism, resulting from the interaction of its genotype with the environment.

Polyhybrid: produced by crossing two or more hybridized parents. These multi-parental crosses contain complex genetic backgrounds, often incorporating alleles from multiple lineages, and are characteristic of most modern commercial cultivars.

Sativa, indica, and hybrid: These denote the common use of the terms describing flowers that have reportedly marked differences in effects (or combination of effects) thought to arise through chemical composition and are unrelated to morphology, geographical origin, or taxonomic description.

Species: A group of organisms capable of interbreeding and producing fertile offspring. In plants, this may include populations that can cross, even if distinct in appearance or ecology. See species concepts.

Strain: An informal term for a named genetic line or type, often with variable genetic consistency. It is commonly used interchangeably with "variety" or as a folk taxonomy. Strain names are not formally described cultivars and should be named with their popularized name without single quotation marks (e.g., Durban Poison and Granddaddy Purple), as these names are not officially recognized through the ICNCP (Pollio 2016).

Terpene: Organic compounds produced by many plants, including cannabis, that contribute to aroma, flavor, and may have biological effects.

Variety: A botanical rank below species, referring to naturally occurring or selectively bred plants with distinct, inherited traits. In agriculture, "variety" and "cultivar" may be used interchangeably, though technically they differ (see below).

References

[1]   Brickell, C.D., Alexander, C., Cubey, J.J., David, J.C., Hoffmann, M.H.A., Leslie, A.C., Malécot, V., Jin, X.B. (2016). International Code of Nomenclature for Cultivated Plants.

[2]   Schwabe, A.L., Vergara, D. The Debated Nomenclature of Cannabis sativa. In: The Cannabis Genome. pp. 1–23. CRC Press.

[3]   Small, E. (2015). Evolution and classification of Cannabis sativa (marijuana, hemp) in relation to human utilization. The botanical review, 81(3), 189–294. https://doi.org/10.1007/s12229-015-9157-3.

[4]   Clarke, R., Merlin, M. (2016). Cannabis: evolution and ethnobotany. Univ of California Press.

[5]   Ren, G., Zhang, X., Li, Y., Ridout, K., Serrano-Serrano, M.L., Yang, Y., Liu, A., Ravikanth, G., Nawaz, M.A., Mumtaz, A.S., Salamin, N. (2021). Large-scale whole-genome resequencing unravels the domestication history of Cannabis sativa. Science advances, 7(29), eabg2286. doi: 10.1126/sciadv. abg2286.

[6]   Balant, M., Vitales, D., Wang, Z., Barina, Z., Fu, L., Gao, T., Garnatje, T., Gras, A., Hayat, M.Q., Oganesian, M., Pellicer, J. (2025). Integrating target capture with whole genome sequencing of recent and natural history collections to explain the phylogeography of wild-growing and cultivated cannabis. Plants, People, Planet, https://doi.org/10.1002/ppp3.70043.

[7]   Sawler, J., Stout, J.M., Gardner, K.M., Hudson, D., Vidmar, J., Butler, L., Page, J.E., Myles, S. (2015). The genetic structure of marijuana and hemp. PloS one, 10(8), e0133292. https://doi.org/10.1371/journal. pone.0133292.

[8]   Schwabe, A.L., Hansen, C.J., Hyslop, R.M., McGlaughlin, M.E. (2021). Comparative genetic structure of Cannabis sativa including federally produced, wild collected, and cultivated samples. Frontiers in Plant Science, 12, 675–770. https://doi.org/10.3389/fpls.2021.675770.

[9]   Vergara, D., Huscher, E.L., Keepers, K.G., Pisupati, R., Schwabe, A.L., McGlaughlin, M.E., Kane, N.C. (2021). Genomic evidence that governmentally produced Cannabis sativa poorly represents genetic variation available in state markets. Frontiers in plant science, 12, 668315. https://doi.org/10.3389/ fpls.2021.668315.

[10]   Halpin-mccormick, A., Heyduk, K., Kantar, M.B., Batora, N.L., Masalia, R.R., Law, K.B., Kuntz, E.J. (2024). Examining population structure across multiple collections of Cannabis. Genetic Resources and Crop Evolution, 71(8), 4705–4722. https://doi.org/10.1007/s10722-024-01928-1.

[11]   Atakan, Z., Bhattacharyya, S., Allen, P., Martín-Santos, R., Crippa, J.A., Borgwardt, 2., Fusar-Poli, P., Seal, M., Sallis, H., Stahl, D., Zuardi, A.W. (2013). Cannabis affects people differently: inter-subject variation in the psychotogenic effects of Δ9-tetrahydrocannabinol: a functional magnetic resonance imaging study with healthy volunteers. Psychological medicine, 43(6), 1255–1267. doi: 10.1017/ S0033291712001924.

[12]   Bidwell, L.C., Ellingson, J.M., Karoly, H.C., YorkWilliams, S.L., Hitchcock, L.N., Tracy, B.L., Klawitter, J., Sempio, C., Bryan, A.D., Hutchison, K.E. (2020). Association of naturalistic administration of cannabis flower and concentrates with intoxication and impairment. JAMA psychiatry, 77(8), 787–796. doi: 10.1001/jamapsychiatry.2020.0927.

[13]   Cloutier, R.M., Calhoun, B.H., Linden-Carmichael, A.N. (2022). Associations of mode of administration on cannabis consumption and subjective intoxication in daily life. Psychology of Addictive Behaviors, 36(1), 67. https://doi.org/10.1037/adb0000726.

[14]   Ayyagari, M.M., Heim, D., Sumnall, H.R., Monk, R.L. (2024). Contextual factors associated with subjective effects of cannabis: A systematic review and meta-analysis. Neuroscience & Biobehavioral Reviews, 164, 105822. https://doi.org/10.1016/j.neubiorev.2024.105822.

[15]   Coyne, J.A., Orr, H.A. (2004). Speciation: a catalogue and critique of species concepts. Philosophy of biology: an anthology, 272–292.

[16]   Baker, R.J., Bradley, R.D. (2006). Speciation in mammals and the genetic species concept. Journal of mammalogy, 87(4), 643–662. https://doi.org/10.1644/06-MAMM-F-038R2.1.

[17] Vilà, C., Maldonado, J.E., Wayne, R.K. (1999). Phylogenetic relationships, evolution, and genetic diversity of the domestic dog. Journal of Heredity, 90(1), 71–77. https://doi.org/10.1093/jhered/90.1.71.

[18] Pollinger, J.P., Earl, D.A., Knowles, J.C., Boyko, A.R., Parker, H., Geffen, E., Pilot, M., Jedrzejewski, W., Jedrzejewska, B., Sidorovich, V., Greco, C. (2011). A genome-wide perspective on the evolutionary history of enigmatic wolf-like canids. Genome research, 21(8), 1294–1305. doi: 10.1101/gr.116301.110.

[19] Gabryś, J., Kij, B., Kochan, J., Bugno-Poniewierska, M. (2021). Interspecific hybrids of animals-in nature, breeding and science–a review. Annals of Animal Science, 21(2), 403–415. doi: 10.2478/aoas-2020-0082.

[20] De Queiroz, K. (2005). Ernst Mayr and the modern concept of species. Proceedings of the National Academy of Sciences, 102(suppl_1), 6600–6607. https://doi.org/10.1073/pnas.0502030102.

[21] Van Valen, L. (1976). Ecological species, multispecies, and oaks. Taxon, 233–239. https://doi.org/10.2307/1219444.

[22] Cracraft, J. (1983). Species concepts and speciation analysis. In: Current ornithology. pp. 159–187. New York, NY: Springer US.

[23] Wheeler, Q.D. (2000). A critique from the Wheeler and Platnick phylogenetic species concept perspective: Problems with alternative concepts of species. Species concepts and phylogenetic theory: a debate, 133–145.

[24] Linnaeus, C. (1753). Species Plantarum. Vol. 2. Stockholm: Laurentii Salvii.

[25] Lamarck, J.B.A.P.D.M.D. (1785). Encyclopédie Méthodique. In: Botanique. Vol. 1. Paris: Panckoucke.

[26] Small, E., Cronquist, A. (1976). A practical and natural taxonomy for Cannabis. Taxon, 405–435. https://doi.org/10.2307/1220524.

[27] United States v. Rothberg (1972). 351 F.Supp. 1115 (E.D.N.Y.), aff'd (1973) 480 F.2d 534 (2d Cir.).

[28] People v. Van Alstyne (1975). 46 Cal.App.3d 900 (Cal. Ct. App., 2d Dist. Div. 3).

[29] Schultes, R.E., Klein, W.M., Plowman, T., Lockwood, T.E. (1974). Cannabis: an example of taxonomic neglect. Botanical Museum Leaflets, Harvard University, 23(9), 337–367. http://www.jstor.org/stable/41762285.

[30] McPartland, J.M., Guy, G.W. (2017). Models of Cannabis taxonomy, cultural bias, and conflicts between scientific and vernacular names. The botanical review, 83(4), 327–381. https://doi.org/10.1007/s12229-017-9187-0.

[31] Turland, N.J., Wiersema, J.H., Barrie, F.R., Gandhi, K.N., Gravendyck, J., Greuter, W., Hawksworth, D.L., Herendeen, P.S., Klopper, R.R., Knapp, S., Kusber, W.H. (2025). International Code of Nomenclature for algae, fungi, and plants (Madrid Code). University of Chicago Press.

[32] SeedFinder. (n.d.). Cultivators Choice Cannabis Strains. Available at: https://en.seedfinder.eu/database/breeder/cultivators-choice, Accessed Sept 10, 2025.

[33] Leafly (2025) Strains. Available at: https://www.leafly.com/strains Accessed October 7 2025)

[34] CannaConnection (2025) Marijuana strains database – CannaConnection. Available at: https://www.cannaconnection.com/strains (Accessed October 7 2025).

[35] Weedmaps (2025) Explore Weed Strains Info, Photos, and Reviews on Weedmaps. Available at: https://weedmaps.com/strains. (Accessed October 7 2025)

[36] ISHS (n.d.). International Code of Nomenclature for Cultivated Plants (ICNCP). International Society for Horticultural Science. Available at: https://www.ishs.org/nomenclature-and-cultivar-registration

[37] Janischevsky, D.E. (1924). Cannabis ruderalis. Proceedings Saratov, 2(2), 14–15.

[38] Schwabe, A.L., McGlaughlin, M.E. (2019). Genetic tools weed out misconceptions of strain reliability in Cannabis sativa: implications for a budding industry. Journal of Cannabis Research, 1(1), 3. https://doi.org/10.1186/s42238-019-0001-1.

[39] Grassa, C.J., Weiblen, G.D., Wenger, J.P., Dabney, C., Poplawski, S.G., Timothy Motley, S., Michael, T.P., Schwartz, C.J. (2021). A new Cannabis genome assembly associates elevated cannabidiol (CBD)

with hemp introgressed into marijuana. New Phytologist, 230(4), 1665–1679. https://doi.org/10.1111/nph.17243.

[40] Lynch, R.C., Padgitt-Cobb, L.K., Garfinkel, A.R., Knaus, B.J., Hartwick, N.T., Allsing, N., Aylward, A., Bentz, P.C., Carey, S.B., Mamerto, A., Kitony, J.K. (2025). Domesticated cannabinoid synthases amid a wild mosaic cannabis pangenome. Nature, 1–10. https://doi.org/10.1038/s41586-025-09065-0.

[41] Zhang, Q., Chen, X., Guo, H., Trindade, L.M., Salentijn, E.M.J., Guo, R., Guo, M., Xu, Y., Yang, M. (2018). Latitudinal Adaptation and Genetic Insights into the Origins of Cannabis sativa L. Frontiers in Plant Science, 9, 1876. https://doi.org/10.3389/fpls.2018.01876.

[42] De meijer, E.P., Bagatta, M., Carboni, A., Crucitti, P., Moliterni, V.M., Ranalli, P., Mandolino, G. (2003). The inheritance of chemical phenotype in Cannabis sativa L. Genetics, 163(1), 335–346. https://doi.org/10.1093/genetics/163.1.335.

[43] Hillig, K.W., Mahlberg, P.G. (2004). A chemotaxonomic analysis of cannabinoid variation in Cannabis (Cannabaceae). American Journal of Botany, 91(6), 966–975. https://doi.org/10.3732/ajb.91.6.966.

[44] Hazekamp, A., Tejkalová, K., Papadimitriou, S. (2016). Cannabis: from cultivar to chemovar II—a metabolomics approach to Cannabis classification. Cannabis and Cannabinoid Research, 1(1), 202–215. https://doi.org/10.1089/can.2016.0017.

[45] Smith, C.J., Vergara, D., Keegan, B., Jikomes, N. (2022). The phytochemical diversity of commercial Cannabis in the United States. PLoS one, 17(5), e0267498. https://doi.org/10.1371/journal.pone.0267498.

[46] Small, E., Beckstead, H.D. (1973). Cannabinoid phenotypes in Cannabis sativa. Nature, 245(5421), 147–148. https://doi.org/10.1038/245147a0.

[47] Fournier, G., Richez-Dumanois, C., Duvezin, J., Mathieu, J.P., Paris, M. (1987). Identification of a new chemotype in Cannabis sativa: cannabigerol-dominant plants, biogenetic and agronomic prospects. Planta Medica, 53(03), 277–280. https://doi.org/10.1055/s-2006-962705.

[48] Mandolino, G., Carboni, A. (2004). Potential of marker assisted selection in hemp genetic improvement. Euphytica, 140, 107–120. https://doi.org/10.1007/s10681-004-4759-6.

[49] Oswald, I.W., Ojeda, M.A., Pobanz, R.J., Koby, K.A., Buchanan, A.J., Del Rosso, J., Guzman, M.A., Martin, T.J. (2021). Identification of a new family of prenylated volatile sulfur compounds in cannabis revealed by comprehensive two-dimensional gas chromatography. ACS omega, 6(47), 31667–31676. https://doi.org/10.1021/acsomega.1c04196.

[50] Plumb, J., Demirel, S., Sackett, J.L., Russo, E.B., Wilson-Poe, A.R. (2022). The nose knows: Aroma, but not THC mediates the subjective effects of smoked and vaporized cannabis flower. Psychoactives, 1(2), 70–86. https://doi.org/10.3390/psychoactives1020008.

[51] Oswald, I.W., Paryani, T.R., Sosa, M.E., Ojeda, M.A., Altenbernd, M.R., Grandy, J.J., Shafer, N.S., Ngo, K., Peat III, J.R., Melshenker, B.G., Skelly, I. (2023). Minor, nonterpenoid volatile compounds drive the aroma differences of exotic cannabis. ACS omega, 8(42), 39203–39216. https://doi.org/10.1021/acsomega.3c04496.

[52] Paryani, T.R., Sosa, M.E., Page, M.F., Martin, T.J., Hearvy, M.V., Ojeda, M.A., Koby, K.A., Grandy, J.J., Melshenker, B.G., Skelly, I., Oswald, I.W. (2024). Nonterpenoid Chemical Diversity of Cannabis Phenotypes predicts differentiated aroma characteristics. ACS omega, 9(26), 28806–28815. https://doi.org/10.1021/acsomega.4c03225.

[53] Toews, D.P., Taylor, S.A., Vallender, R., Brelsford, A., Butcher, B.G., Messer, P.W., Lovette, I.J. (2016). Plumage genes and little else distinguish the genomes of hybridizing warblers. Current Biology, 26(17), 2313–2318. doi: 10.1016/j.cub.2016.06.034.

[54] Van Treuren, R., Kemp, H., Ernsting, G., Jongejans, B., Houtman, H., Visser, L. (2010). Microsatellite genotyping of apple (Malus × domestica Borkh.) genetic resources in the Netherlands: Application in collection management and variety identification. Genetic Resources and Crop Evolution, 57(6), 853–865. https://doi.org/10.1007/s10722-009-9525-0.

[55] Evans, K.M., Patocchi, A., Rezzonico, F., Mathis, F., Durel, C.E., Fernandez-Fernandez, F., Boudichevskaia, A., Dunemann, F., Stankiewicz-Kosyl, M., Gianfranceschi, L., Komjanc, M. (2011). Genotyping of pedigreed apple breeding material with a genome-covering set of SSRs: trueness-to-type of cultivars and their parentages. Molecular Breeding, 28(4), 535–547. https://doi.org/10.1007/s11032-010-9502-5.

[56] Cabezas, J.A., Ibáñez, J., Lijavetzky, D., Vélez, D., Bravo, G., Rodríguez, V., Carreño, I., Jermakow, A.M., Carreño, J., Ruiz-García, L., Thomas, M.R. (2011). A 48 SNP set for grapevine cultivar identification. BMC plant biology, 11(1), 153. https://doi.org/10.1186/1471-2229-11-153.

[57] Sebastiani, L., Busconi, M. (2017). Recent developments in olive (Olea europaea L.) genetics and genomics: applications in taxonomy, varietal identification, traceability and breeding. Plant cell reports, 36(9), 1345–1360. https://doi.org/10.1007/s00299-017-2145-9.

[58] Larsen, B., Gardner, K., Pedersen, C., Ørgaard, M., Migicovsky, Z., Myles, S., Toldam-Andersen, T.B. (2018). Population structure, relatedness and ploidy levels in an apple gene bank revealed through genotyping-by-sequencing. PLoS One, 13(8), e0201889. https://doi.org/10.1371/journal.pone.0201889.

[59] Jodrey, K., personal communication, 2025.

[60] Wilson, J. (2022). Curious About Cannabis: A Scientific Introduction to a Controversial Plant. 3rd edn. ISBN 978–0-9985728–8-8.

Bianca Schnarr

Parallel cannabinoid chemistries in nature: non-cannabis sources and endocannabinoid modulators

Introduction: *Cannabis*, a chemical pharmacopeia

Cannabis is a dynamic and chemically complex plant, biosynthesizing over 550 known compounds like flavonoids, terpenes, and fatty acids. These compounds are found in varying quantities throughout the plant, from the roots to the seed. It has been estimated that at least 120 of these molecules are cannabinoids. These cannabinoids are also known as phytocannabinoids, meaning they are synthesized by plants; the prefix "phyto" means "plant" in Greek. Phytocannabinoids are a complex class of compounds, with the most famed being delta-9-tetrahydrocannabinol (Δ9-THC), *Cannabis'* primary psychoactive constituent. Additional cannabinoids with growing popularity are cannabidiol (CBD), cannabichromene (CBC), cannabinol (CBN), cannabigerol (CBG), and cannabivarin [1]. Cannabinoids like THC and CBN are known for their ability to exert psychoactivity, though the vast majority of cannabinoids have no psychoactive effect at all; rather they contribute to the modulation of a milieu of other biochemical processes within the body. This class of compounds was once believed to exist exclusively within *Cannabis* spp., of which there are three distinct species: *Cannabis sativa*, *Cannabis indica*, and *Cannabis ruderalis*. Revelations in phytochemistry over the past few decades have unearthed an array of molecules with similar structures and potential pharmacological effects, demonstrating that phytocannabinoids are more widespread than previously imagined.

To understand the rich world of cannabinoids both within and outside of *Cannabis* sp., we must first explore what it means to be a cannabinoid and how these compounds exert their effect on the human body.

The endocannabinoid system

The endocannabinoid system (ECS) is a modulatory network of receptors and endogenous ligands found across the animal kingdom [2]. The ECS regulates a host of essential processes throughout the body and mind. This system of G protein-coupled receptors (GPCRs) is one of the most widespread in the human body. This system helps the body maintain homeostasis by modulating stress response, reward signaling, gut mobility, intestinal inflammation, and neuroplasticity [3, 4].

© 2026 Walter de Gruyter GmbH, Berlin | https://doi.org/10.1515/9783111476162-002

The best-characterized receptors in the ECS are cannabinoid receptor 1 (CB1 receptor) and cannabinoid receptor 2 (CB2 receptor) [2]. Both are presynaptic inhibitory receptors, meaning they inhibit other receptors downstream, such as glutamate and gamma-aminobutyric acid (GABA), which are the central nervous system's (CNS) major excitatory and inhibitory receptors, respectively [1–3].

The CB1 receptor is the most abundant GPCR in the CNS, where it is predominantly expressed. In the CNS, the CB1 receptor has been shown to regulate memory, appetite, emotion, coordination, fear response, and decrease inflammation. The brain regions with the highest density of CB1 receptors are the hippocampus, hypothalamus, amygdala, cerebellum, and nucleus accumbens. CB1 receptor expression is not limited to the brain; this receptor type is also found in the gut. Research suggests the CB1 receptors in the gut promote feeding through stimulating the inhibition of cholecystokinin (CKK) induced satiation. CKK is a hormone released from the small intestine that triggers one to recognize they are full. Despite the milieu of benefits the CB1 receptor offers, overstimulation of this receptor type has been shown to increase inflammation [1, 3].

The CB2 receptor is found primarily in the peripheral nervous system, where it is expressed in a variety of cell types [1]. The highest density of CB2 receptors is expressed in the gastrointestinal tract, spleen, lymphocytes, and bone marrow. A growing body of evidence demonstrates that activation of the CB2 receptor decreases cellular inflammation. The result of this decreased inflammatory response includes cardioprotective effects, immune support, and enhanced functioning [3]. Like the CB1 receptor, overstimulation of the CB2 receptor can lead to counteractive effects such as decreased immune function [1].

The ECS is not limited to the CB1 and CB2 receptors. A growing number of receptor types are believed to be a part of the ECS. These include transient receptor potential vanilloid receptors (TRPVRs), which regulate temperature sensation, smell, taste, vision, pressure, and pain perception. Peroxisome proliferator-activated receptor γ (PPARγ) regulates the expression of transcription factors, thereby controlling which genetic sequences are activated or repressed. The final subclass of receptors currently believed to be involved in the ECS are orphan GPCRs, such as GPR55, which has been found to increase signaling cascades through the potentiation of intracellular calcium, and GPR18, which has been shown to modulate inflammatory and nociceptive responses in the body [1, 5–7].

Cannabinoids

Endocannabinoids

Endocannabinoids (ECBs) are naturally occurring signaling molecules found throughout the animal kingdom. These compounds are the body's natural way of utilizing the ECS without external supplementation. Anandamide (AEA) and 2-arachidonoylglycerol (2-AG) are the most well-studied ECBs in terms of pharmacologic and metabolic activity. AEA is a partial agonist of the CB1 receptor with high affinity while maintaining virtually no activity at the CB2 receptor. In biochemistry, the word affinity refers to the strength with which a compound, either made within the body (endogenous) or introduced into the body (exogenous), can bind to a particular receptor or enzyme (see Box 2.1).

> **Box 2.1: Affinities explained**
> Affinity refers to how strongly a compound binds to a specific receptor or enzyme in the body. The more strongly a compound binds to a particular receptor, the more likely it is to trigger a cellular response. In the case of cannabinoids, two major receptors are involved: **CB1** (primarily localized in the central nervous system, i.e., the brain and spinal cord) and **CB2** (primarily localized in the immune system in the periphery and brain). Affinity values are defined using the symbol K_i (for affinity constant). The amount of a compound that is required to trigger a response in a particular receptor or enzyme is typically quantified in nanomoles (nM). A lower Ki means that less of the compound is required to trigger a response or that the compound binds more tightly to the receptor and is more likely to activate or block its effects. Therefore, a **low K_i value** indicates **high affinity**, while a **high K_i value** indicates **low affinity**. By comparing the affinities of different cannabinoids to Δ9-THC (the primary psychoactive compound in cannabis), we can get a sense of how other cannabinoids might interact with the body. For example, if a cannabinoid has a **higher K_i value** for CB1 than Δ9-THC, it binds less strongly and may be **less psychoactive**. If it has a **lower K_i value** for CB2 than Δ9-THC, it binds more strongly and may have **greater potential for immune or anti-inflammatory effects** [1, 2, 8].

A cell can either be activated or inhibited when bound to a compound, depending on that compound's functionality. Agonism is when the binding of a compound activates a particular receptor type. For GPCRs, this activation triggers a conformational change that allows the receptor to interact with G proteins. This interaction triggers an intracellular signaling cascade, causing a biological response. For example, 2-AG is an agonist of the CB1 receptor. When it is bound, it alters the receptor's conformation, allowing it to interact with certain G proteins, which in turn decrease adenylate cyclase and reduce cyclic adenosine monophosphate (cAMP), resulting in the biological response of suppressing synaptic signaling [2, 8]. Antagonism is when a compound binding blocks the activation of a particular receptor through binding to the same site (known as the orthosteric site) as an agonist would normally bind. AEA has also been found to interact with receptors in other systems, like the serotonin and muscarinic receptors, and glycine and nicotinic receptors in vitro. On the other hand, 2-AG is a full agonist at both the CB1 and the CB2 receptors [3, 8]. To put the difference between some exogenous cannabinoids and endogenous cannabinoids into perspective, THC is a partial CB1 receptor agonist, while 2-AG is a full agonist [2]. This means that 2-AG can activate the CB1 receptor to its full potential, whereas THC, even at high concentrations, produces a maximal response that is lower than that of 2-AG. Despite this, THC elicits a response at lower concentrations than 2-AG due to its higher affinity for the CB1 receptor, even though it is only a partial agonist. [1, 2].

Another compound functional class one should be familiar with is allosteric modulators. These compounds influence a receptor or enzyme activity by binding to a site other than the orthosteric site. There are positive allosteric modulators, which improve protein activity, and negative allosteric modulators, which decrease the protein's response [2]. The final functional class to discuss is inverse agonists; they are like antagonists as they both block agonist access to the binding site, though inverse agonists uniquely decrease activity below basal signaling [1]. The relative affinities and functionalities of selected ECBs and cannabinoids are shown in Table 1.

Table 1: Receptor affinities for select endocannabinoids and phytocannabinoids.

Agent	CB1 affinity and functionality (K_i^a)	CB2 affinity and functionality (K_i^a)
(−)-Delta-9-*trans*-tetrahydrocannabinols (Δ9-THCs)[b]	25.1 (reference) [1, 10] (partial agonist)	35.2 (reference) [1, 10] (partial agonist)
2-Arachidonoylglycerol (2-AG)	3,243.61 (−) [10]	1,900 (−) [10]
Anandamide (AEA)	239.2 (−) [10]	439.5 (−) [10]
Cannabidiol (CBD)	304 (allosteric site) [11] (negative allosteric modulator)	2,860 (−) [1] (antagonist)
Cannabigerol (CBG)	440−1,045 (−) [1, 9] (weak agonist)	153.4−1,225 (−) [1, 9] (partial agonist)
Cannabinol (CBN)	525.3 (−) [1, 10] (agonist)	168.2 (−) [1, 10] (inverse agonist)

[a]Values are K_i binding affinities in nM for human receptors.
[b](+) indicates higher affinity than Δ9-THC; (−) indicates lower affinity than Δ9-THC.

The most well-known metabolizer of AEA is fatty acid amide hydrolase (FAAH), and for 2-AG, it is monoacylglycerol, and both ligands can also be metabolized by the liver's cytochrome P450 (CYP) class of enzymes [2]. These major modes of signal termination are utilized when exogenous cannabinoids are ingested.

Endocannabinoid-like (ECL) compounds

ECLs, also known as cannabimimetics, are a class of endogenous and exogenous compounds that influence the ECS without directly activating CB1 and CB2 [7]. These compounds typically propagate their action through inhibition of ECB metabolism or activate ECB receptors other than CB1 and CB2 [7]. Additionally, they may interact with TRPVRs, PPARγ, and orphan GPCRs like GPR55 and GPR18 [1, 7]. These compounds regulate a range of essential bodily functions like appetite, temperature perception, energy metabolism, and the immune system [7]. ECL compounds are believed to be found in *Olea europaea*, "green olive" oil and *Theobroma cacao*, "chocolate" [7, 12].

Phytocannabinoids in *Cannabis*

There have been over 113 different cannabinoids isolated from *C. sativa*, which are subcategorized into the following distinct categories based on structure: CBGs, CBCs, CBDs, Δ9-THCs, (−)-Delta-8-*trans*-tetrahydrocannabinols (Δ8-THCs), cannabicyclols (CBLs), cannabielsoins (CBEs), CBNs, cannabinodiols (CBNDs), cannabitriols (CBTs), and the miscellaneous cannabinoids [5]. Phytocannabinoids are secondary metabolites, meaning they are not essential for the plant's metabolism, growth, and reproduction, though these compounds do improve an organism's chances of survival in response to environmental stress. Secondary metabolites are a class of compounds that encompass a variety of molecules such as cannabinoids, terpenes, and alkaloids. These chemicals defend plants against environmental stresses like UV rays, microbial and viral pathogens, predatory insects, and cold temperatures. Cannabinoids and terpenes in *C. sativa* are generated in the glandular trichomes. The highest concentration of trichomes is found on the mature female flower, with lower expression levels across other aerial parts of the plant, like the leaves [5].

There are different subcategories of plant-derived compounds that fall under the umbrella of "cannabinoid": structural phytocannabinoids and pharmacological phytocannabinoids.

Structural phytocannabinoids

Structural phytocannabinoids are defined by their chemical arrangement. To fit within this category, they must include 21 aromatic hydrocarbons, at least 1 oxygen, contain a hydrophobic alkyl chain, and cannot contain nitrogen (Figure 1) [2]. Figure 1 shows the chemical structures of the basic phytocannabinoids and key phytocannabinoids found in *C. sativa*, *Helichrysum umbraculigerum*, *Radula perrottetii*, *R. marginata*, *Rhododendron dauricum*, and *Rhododendron anthopogonides*. These include CBGA, CBD, CBC,

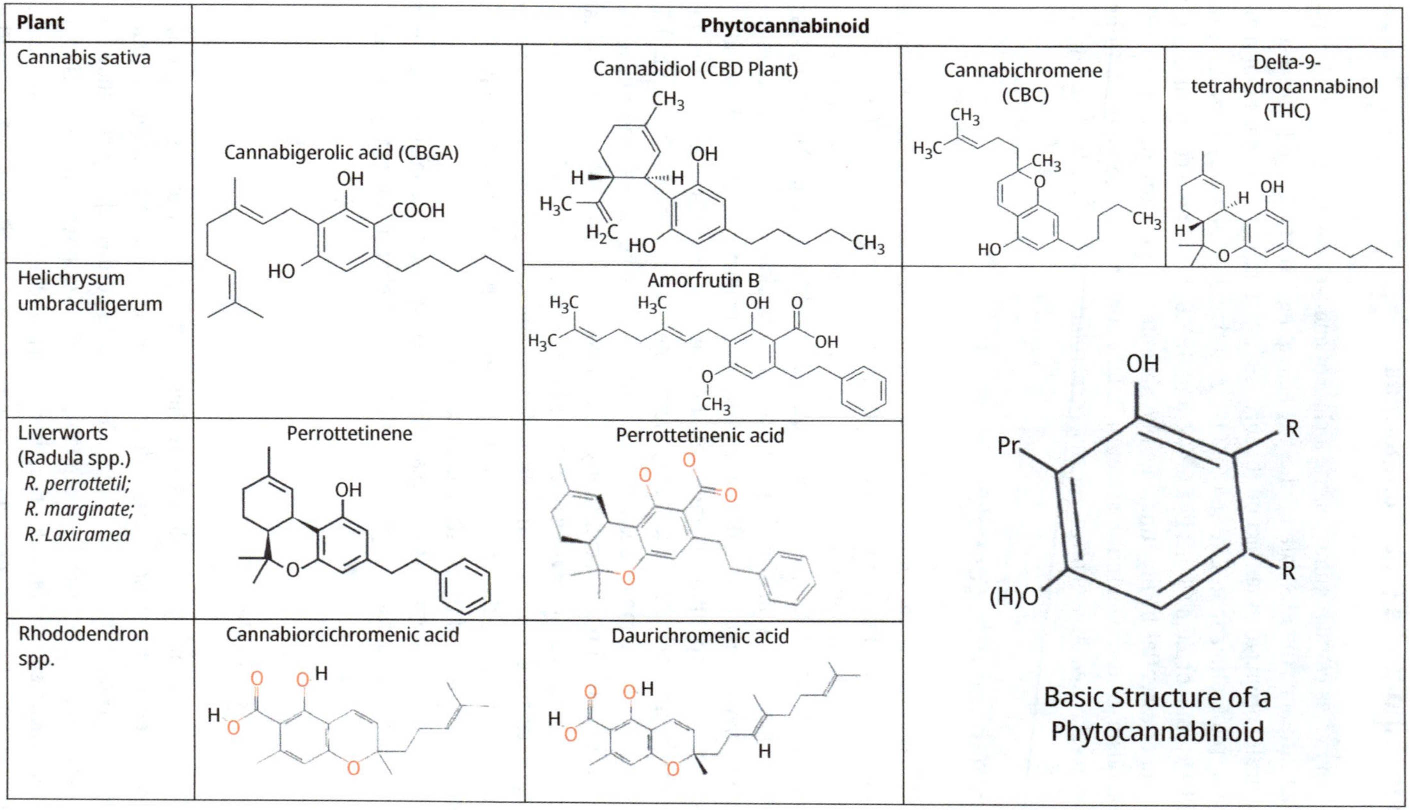

Plant	Phytocannabinoid			
Cannabis sativa	Cannabigerolic acid (CBGA)	Cannabidiol (CBD Plant)	Cannabichromene (CBC)	Delta-9-tetrahydrocannabinol (THC)
Helichrysum umbraculigerum		Amorfrutin B		
Liverworts (Radula spp.) *R. perrottetil; R. marginate; R. Laxiramea*	Perrottetinene	Perrottetinenic acid		
Rhododendron spp.	Cannabiorcichromenic acid	Daurichromenic acid	Basic Structure of a Phytocannabinoid	

Figure 1: Structural phytocannabinoids found inside and outside *Cannabis* spp.

THC, amorfrutin B, perrottetinene, perrottetinenic acid, cannabiorcichromenic acid (CBCA), and daurichromenic acid (DCA).

Pharmacological phytocannabinoids

Pharmacological cannabinoids are compounds that directly interact with CB1 or CB2 receptors. They can also demonstrate activity at the TRP, PPARγ, and orphan GPCRs [5]. To gain a better understanding of what it means to be a pharmacological cannabinoid, the known biochemical effects of *C. sativa*'s most well-documented pharmacologically active phytocannabinoid, Δ9-THC, will be explored.

To begin, the effects elicited by Δ9-THC's binding to the CB1 and CB2 receptors will be further elucidated. The psychological effects associated with this molecule are due to its activation of the CB1 receptor. The effects associated with this molecule include psychoactivity, analgesia, relaxation, antiemetic, and antispasmodic properties [1, 5]. Δ9-THC also binds to an array of other receptors, which all contribute to its effects. Δ9-THC is a partial agonist at the CB2 receptor; it activates GPR55 and GPR18, all of which have been shown in animal models to produce a potent anti-inflammatory and antinociceptive effect. Δ9-THC is also a positive allosteric modulator of the delta and mu-opioid receptors, increasing the potency of opioids bound to the receptor. Finally, Δ9-THC is an agonist of TRPV channels 2, 3, and 4 [1]. All these interactions contribute to Δ9-THCs dynamic effects and therapeutic potential while demonstrating that one cannot entirely attribute this substance's effects to a single receptor.

When using the whole plant, fungi, or tissue formulations of any natural product, the contribution of any major metabolites should be considered when evaluating the cause of the therapeutic effect. *Cannabis*, in addition to all the plants to be explored in this chapter, is subject to a phenomenon known as the entourage effect. The entourage effect, in short, details that natural products hold a pharmacopeia of compounds within a given plant, animal, bacteria, or fungi and all the compounds in these natural products that are expressed in levels high enough to exert a biochemical effect, hold the potential to influence the downstream pharmacological and phenomenological effects in those who consume it [13]. Moreover, as demonstrated with Δ9-THC, the action of a single compound is not limited to one specific receptor subtype; therefore, all interactions should be considered when evaluating medical potential.

After looking at what it means to be a cannabinoid, endogenously, structurally, and pharmacologically, exploration of the plants that express these compounds outside of the *Cannabis* genus can finally begin.

Plants that produce *Cannabis*-type cannabinoids

Helichrysum umbraculigerum

Helichrysum umbraculigerum is the only other plant outside of *C. sativa* that produces *Cannabis*-type cannabinoids. *Cannabis*-type cannabinoids are, in the simplest terms, the same cannabinoids found in *Cannabis* spp. The presence of cannabigerolic acid or CBGA, has been detected in *H. umbraculigerum* and *C. sativa* (see Figure 1). Quantities of CBGA in *H. umbraculigerum* have been reported at levels up to 4.3%. To put this in context, this is greater than the maximum CBGA or CBG content observed across 320 *Cannabis* chemotypes [14–16]. When CBGA is paired with endogenous enzymatic catalysts called synthases in *C. sativa* CBGA serves as the direct precursor to tetrahydrocannabinolic acid (THCA), cannabidiolic acid, and cannabichromenic acid (CBCA). When these compounds are deacidified, also known as decarboxylated, through high temperature via combustion and vaporization, they become CBG, THC, CBD, and CBC, respectively. [17]. Like *Cannabis*, the biosynthesis and storage of CBGA in *H. umbraculigerum* occur in the glandular trichomes of the plant's aerial parts. In contrast, *H. umbraculigerum* primarily produces cannabinoids in the leaves rather than in the mature flower. The production of identical cannabinoids in two phylogenetically different species is fascinating, as it demonstrates independent yet parallel evolution, also known as convergent evolution [14].

CBG is a non-psychotropic phytocannabinoid that has a low affinity for the CB1 and CB2 receptors and has the capacity to inhibit AEA uptake. It also blocks the 5-HT1A receptor. CBG is a transient receptor potential melastatin 8 antagonist and transient receptor potential ankyrin 1 (TRPA1) agonist; both receptors play a major role in one's body's detection of cold temperatures and certain irritants [1, 5]. *Helichrysum umbraculigerum* also produces benzylic analogs of CBG (bibenzyl-CBG). Like CBG, these analogs display low affinity for CB1 and CB2 receptors. They have high affinity for the ionotropic receptors TRPV 1–4 and TRPA 1, with some displaying increased affinity towards TRPM 8 [5].

Helichrysum species are omnipresent in South African medicine, customarily used to treat common illnesses such as respiratory and skin infections [5, 18]. Ritualistic uses of *H. umbraculigerum* have also been reported, primarily for fumigation, air purification, or as an incense to invoke the goodwill of the ancestors [6, 14, 15]. Ethnobotanical studies have found *H. umbraculigerm*, and related species, have a history of psychotropic use, suggesting a chemical within the plant acts on the CNS. Subsequent binding studies using *H. umbraculigerm* extract exhibited GABA-receptor activity, though the primary compound responsible for this activity is yet to be elucidated [5, 19]. Livestock grazing material is another traditional use of *H. umbraculigerm*.

In vitro screening methods have revealed that this species of *Helichrysum* also holds antibacterial and antifungal activity [15]. There are a few other compounds within *H. umbraculigerm* that require improved characterization to have a better understanding of the impact they have on humans pharmacologically and clinically. Additional secondary metabolites this species produces, which are structurally characterized as cannabinoids, include amorfrutin B, as well as polyphenols and flavonoids like prenylchalcones and prenylflavanones [14].

β-caryophyllene-containing plants

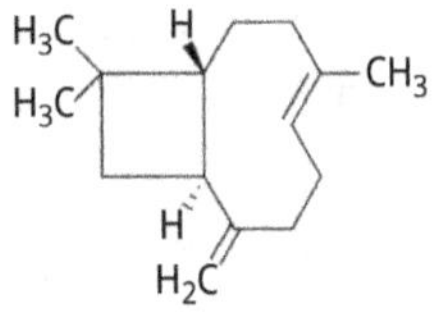

β-caryophyllene

Figure 2: Chemical structure of β-caryophyllene. (structure sourced from Wikipedia).

Hundreds of plants express substantial concentrations (E)-β-caryophyllene (BCP), *C. sativa* included [20]. BCP is a bicyclic sesquiterpene hydrocarbon that demonstrates selective full agonist properties at the CB2 receptor (see Figure 2). BCP is one of the only compounds that has been validated to have pharmacological activity on the ECS outside of the 113 different cannabinoids found in *Cannabis* [12]. This notion has been validated via pre-clinical models with CB2 KO mice by Gertsch et al. [12]. In addition to CB2 activation, BCP also activates peroxisome proliferator-activated receptor-α (PPARα) [12, 21]. Activation of the CB2 receptor is highly implicated in immune response, analgesia, and inflammation. The activation of the CB2 receptor pathway entails the suppression of cytokine release from immune cells, which decreases the inflammatory response [21]. BCP has demonstrated therapeutic benefits for a range of conditions, including cancer, chronic pain, and inflammation. BCP has also been approved by both the Food and Drug Administration and the European Food Safety Authority as a flavoring agent, which can be used in cosmetics and food additives. A 2020 database search of botanical sources BCP with a percentage greater than 10% amassed a list containing 295 plants belonging to 51 plant families with samples collected from 56 countries worldwide [21]. A few of these plants will be outlined, highlighting those with the highest BCP percentage and some of the richest histories of use.

Copaifera langsdorffii

Copaifera langsdorffii, popularly known ascopaiba, podoi, or and miracle tree, produces oleoresins (resin oil) that can contain up to 72% BCP [20, 22]. It belongs to the plant family Fabaceae and is commonly used in South American traditional medicine.

Countries with the highest propensity for use include Brazil and Bolivia. *Copaifera langsdorffi* is one of the most widely used medicinal trees in the Amazon, which exemplifies how it acquired the colloquial distinction as the "miracle tree" [20]. There are over 90 health problems that *C. langsdorffi* has been used to treat. The resin oil is used most frequently for infection, inflammation, rheumatic pain, gastritis, and non-specified healing [22]. The methods of administration and indications of use for *C. langsdorffi* resin oil vary based on region. For example, in Floriano-Piauí, it is commonly made into syrups; in this form, it is used to treat burns and throat irritation. While in Claro dos Poções-Minas Gerais, the typical formulation is an aqueous herbal infusion, employed for arthritis, rheumatism, bronchitis, pain in lower limbs, and amelioration of the flu [22].

Piper nigrum

Piper nigrum (Figure 3), or "black pepper" seeds, contain a high concentration of BCP, up to 45.3% [20, 23]. *Piper nigrum* is a perennial woody aromatic climber vine native to India, which is best cultivated in hot and moist climates [23]. It has been one of the largest contributors to the global spice market for centuries; in 2024, its global market share accounted for USD4.5 billion. The seed of *P. nigrum* is the most commonly used part. It is a commonplace flavor enhancer and food preserver in Western culture. In India, Ayurvedic medicine heavily employs *P. nigrum* seeds. Clinical uses of black pepper include menstrual disorders such as hypomenorrhea and dysmenorrhea, and ear, nose, and throat disorders such as cough, sinusitis, throat pain, throat infection, and earache [23]. *Piper nigrum* is also used in veterinary medicine for gastrointestinal disorders of livestock. The most common preparations are grinding the seed into a pill, tablet, or paste.

A systematic review found the dominant terpenes in *P. nigrum* to be: BCP (14.7–52.5%), 3-carene (0.8–21.1%), limonene (4.4–18.7%), α-pinene (1.5–7.3%), and β-pinene (1.7–9.4%) [23]. Additionally, compounds found in the dried seeds which contribute to therapeutic properties are: (2E, 4E, 8Z)-*N*-(isobutyl) eicosatrienamide, pellitorine, pipercide, piperine, stigmastanol, stigmasterol, decurrenal, stigmasterol 3-*O*-β-D-glucopyranoside, and 5,10(20)-cadinen-4-ol pipgulzarine, pipzorine, and piptahsine [23]. There is a large body of evidence supporting the notion that *P. nigrum* L. and *Curcuma longa* L. (turmeric) support one another's therapeutic benefits synergistically. This evidence includes in vitro and preclinical rodent studies on the anticancer, antibacterial, antinociceptive, antimalarial, anti-inflammatory, and antioxidant effects of this combination [24–27].

Figure 3: Left: *Piper nigrum* L. plant photographed at Olbrich Botanical Gardens located in Madison, Wisconsin. Right: *Piper nigrum* L. seeds.

Clove

Syzygium aromaticum or "Clove" is another well-known culinary spice with significant levels of BCP [20]. Clove is native to Indonesia and belongs to the largest genus of flowering plants in the world [28]. Traditionally, *S. aromaticum* has been used as an analgesic for toothaches, a digestive aid, and as an anthelmintic [20]. Assays using clove extract have offered a breadth of evidence for antimicrobial, antibacterial, and antifungal activity. Globally, clove is recognized for its unopened flower bud, which is used as a spice. This unopened flower bud is the organ that produces the highest concentration of BCP in the plant [28]. The concentration of BCP in essential oil extracts of clove can reach levels over 27%. This plant also produces high levels of other terpenoids, including eugenol, a compound known for its high antioxidant capacity [20].

Echinacea spp.

Within the plant family of Asteraceae, the largest flowering plant family known to man, lies the genus of perennials known as *Echinacea*. *Echinacea* species are grown all over the world in places like the United States, Canada, and Europe for their aesthetic and medicinal value. *Echinacea* has been used traditionally as an immunomodulator. Documentation of immunomodulation dates back to 1913, when it was first used for the remediation of tuberculosis. It is also used for wound healing, sore throats, and various other bacterial infections [29]. *Echinacea purpurea* (Figure 4) has

Figure 4: *Echinacea purpurea* flowering.

demonstrated antiviral activity by interfering with a virus's ability to bind with host cells. Research has also shown that certain *N*-alkylamides from *E. purpurea* and *Echinacea Angustifolia* selectively bind to the human CB2 receptor with a K_i value of < 100 nM as a partial agonist. These compounds have also been found to inhibit AEA transport and partially inhibit FAAH, as well as bind to PPARγ [1]. In vitro anti-inflammatory studies conducted by Olah et al. in 2017 demonstrated that small nano-molar concentrations of *Echinacea's N*-alkylamides have robust anti-inflammatory effects like the endogenous ECB AEA. This study also presented clinical data showing this extract was able to alleviate clinical symptomology associated with atopic eczema [30].

Brassica spp

The *Brassica* genus is home to many well-known vegetables like kale, broccoli, and collard greens. This genus is also home to a lesser-known chemical compound by the name of DIM (3,3-diindolylmethane). DIM is derived from the conversion of indole-3-carbinol (I3C), which occurs endogenously in acidic environments such as the human stomach [31]. DIM has been found to exhibit hepatoprotective, antioxidant, and anti-cancer properties. DIM's anticancer effects are associated with its ability to reduce inflammation, induce apoptosis, and reduce proliferation and metastasis of tumor cells. This compound is particularly promising in the fight against hormone cancers like prostate cancer. This potential stems from DIM's identity as the first androgen receptor antagonist derived from plants. DIM has additionally demonstrated efficacy on the ECS through activation of both the CB1 and CB2 receptors [31]. Interestingly, preclinical studies indicate that high CB1/CB2 expression in a patient's blood plasma, in addition to elevated levels of endogenous ECB metabolizers, correlates with poorer prognosis in prostate cancer, suggesting the ECS plays a role in prostate cancer cell proliferation. Tucci et al. [31] were the first to describe the effects of cells endogenously expressing CB2 receptors and their associated pathways. Regarding the CB1 receptor, DIM displays slight inverse agonist activity. It has exhibited partial agonist with an affinity or K_i value of approximately 1 μM at the CB2 receptor, though this has not been validated in rodent models to date [1].

Figure 5: *Daucus carota* L.

Apiaceae spp

Falcarinol is a widespread compound found through the plant family Apiaceae. Plants that contain this compound are vegetables found in households throughout the world like carrots (*Daucus carota* L.) (see Figure 5) and celery (*Apium graveolens*), herbs like parsley (*Petroselinum crispum*), as well as *Panax ginseng*, the roots of which have been used in traditional Chinese medicine (TCM) for hundreds of years [1, 13]. Falcarinol has exhibited significant binding with both the CB1 and CB2 receptors, though it shows favorability for the CB1 receptor as an inverse agonist with a binding affinity K_i value of <1 μM. This antagonism potentiates an inflammatory response in human skin [1]. Gonçalves et al. purport that daily dietary supplementation with falcarinol can regulate the gut-microbiota and decrease the number of neoplastic lesions and polyp growth rate in the colon of mice with azoxymethane-induced tumors [13].

Ruta graveolens and *Ruta angustifolia*

Ruta graveolens, also known as common Rue, is a perennial plant native to the Mediterranean. Rue has been employed medicinally by Europeans for over a thousand years and even cited by some of the greatest minds to come out of the region, such as Hippocrates. Traditional uses include rheumatism, dermatitis, pain, and a variety of other inflammatory conditions [32]. The compound in this plant that exhibits cannabinoid activity is Rutamarin. It is found in both *R. graveolens* and *Ruta angustifolia* [33]. Rutamarin has been shown to display selective affinity for the CB2 receptor with a K_i value of <10 μM [1]. Research on the therapeutic potential of Rutamarin has identified promise in this compound's anticancer properties. Rutamarin's cancer defense potential is believed to occur through CB2 receptor-mediated inhibition of the cAMP pathway, thereby inducing apoptosis [33]. To date, this has only been exhibited in cell models, and further research is required to determine the effect of this compound on more complex cell systems and living model organisms.

Chemical cannabinoids

The term cannabinoid generally refers to molecules with a characteristic chemical structure. These are meroterpenoids with a resorcinol core typically decorated with a para-positioned isoprenyl, alkyl, or aralkyl side chain. Another name used for this class of molecules is bibenzyls [5]. Meroterpenoids are natural products that are partially generated by both the terpenoid and polyketide pathways [34]. In contrast to pharmacological phytocannabinoids, these compounds do not necessitate biochemical activity in the animal ECS to fit the distinction of cannabinoid. They can be found in flowering plants, liverworts, and fungi [5].

Liverworts

Liverworts are a subset of small, flowerless, spore-producing, nonvascular plants that grow in moist, shady areas. One of the most interesting facts about this genus is that they are the first plants that migrated from aquatic environments to land. Liverworts have demonstrated a dynamic ability to biosynthesize over 2,200 known secondary metabolites such as indole alkaloids, terpenoids, and flavonoids. These compounds are accumulated in single-membrane-bound organelles, known as oil bodies. *Radula perrottetii*, *Radula marginata*, and *Radula laxiramea* are all species that contain at least one compound with a bibenzyl backbone, qualifying them as phytocannabinoids (see Figure 1). The compounds with the greatest attention are the bibenzyl analog of Δ9-THC, *cis*-perrottetinene. Structurally, it resembles Δ9-THC, distinguishing itself with an aromatic instead of a pentyl side chain [5, 35, 36] (see Figure 1).

In 2018, Chicca et al. published first in vivo studies in mice models to determine the cannabimimetic and biochemical effects, in addition to in vitro structure-activity relationship assays for the CB1R and CB2 of (−)-*cis*-perrottetinene and (−)-*trans*-perrottetinene [37]. It was demonstrated that, like THC, both isomers of perrottetinene interact with the CB1 and CB2 receptors selectively as partial agonists [36]. In these studies, (–)-*cis*-perrottetinene exhibited psychoactive properties as a CB1 agonist, inducing hypothermia, catalepsy, hypo-locomotion, and analgesia, behaviors strongly associated with psychologically altering effects [5]. Research into the clinical properties of perrottetinenes has only recently begun, and more investigation is required to understand its effects and therapeutic potential.

Rhododendron

Phytochemical research on *Rhododendron* species, *R. dauricum* and *R. anthopogonoides*, has determined that both produce meroterpenoids with a cannabinoid backbone (see Figure 1) [5, 14]. These compounds include DCA, rhododaurichromenic acid, anthopogochromenic acid (CBL-type), and CBCA-type from *R. anthopogonoides* [5]. These compounds are associated with anticancer, antimicrobial, anti-inflammatory, antithrombotic, and antipsychotic activities, and show very low toxicity in humans. DCA and rhododaurichromenic acid have been shown to exhibit anti-HIV activity. Anthopogocyclolic acid, anthopogochromenic acid, and CBCA have been shown to inhibit histamine release, therefore producing antiallergic effects [5]. CBC, the decarboxylated form of CBCA, has little to no effect on the CB1 and CB2 receptors.

Rhododendron anthopogonoides is native to China and Tibet. The plant parts with the highest concentration of ECL include the flowers and leaves. *Rhododendron anthopogonoides* is used in TCM for removing body heat, body detoxification, cough, asthma, swelling, and coronary heart disease, and as an anti-inflammatory and smooth muscle relaxant. *Rhododendron dauricum* has long-standing historical use and maintains current uses in TCM in the treatment of respiratory conditions like bronchitis, asthma, and coughs. Additional research suggests that *R. dauricum* holds antioxidant capabilities [38]. Like many of the other newly discovered cannabinoids, there is still little known about the function of cannabinoids in *Rhododendron* species, though based on historical use and what is known thus far, the therapeutic potential is promising [5].

Endocannabinoid-like compounds

Theobroma cacao

The *T. cacao* tree (cacao) (Figure 6) produces one of the most well-known products in the world, i.e., chocolate [12]. The chemical constituents often associated with cacao's biological effects are the alkaloids, theobromine, and caffeine, which are known to release endogenous opioids called endorphins, which produce a stimulating effect. Additional chemicals commonly associated with this plant include its phenolics like cyanidin, (+)-catechin, and (–)-epicatechin. Phenolics are praised for their antioxidant properties, anti-inflammatory effect, and cardioprotective action [39]. Lesser-known constituents of this plant extract are its lipid-derived compounds N-linoleoylethanolamide (N-LEA) and N-oleoylethanolamide (N-OEA), both of which act as ECL compounds, halting the breakdown of ECB AEA by inhibiting metabolic enzyme FAAH [12]. Some hypothesize that N-linoleoyl ethanolamide and N-oleoyl ethanolamide work cooperatively with other phy-

Figure 6: *Theobroma cacao* tree actively flowering photographed at Olbrich Botanical Gardens located in Madison, Wisconsin.

tochemicals in *T. cacao*, like caffeine or theobromine, facilitating acute euphoria following consumption [40].

Traditionally, cacao has been employed across Mesoamerica since as early as 600 BC, where it continues to be revered for its intoxicating and aphrodisiac properties [41]. Some areas of this region, like modern-day Guatemala, categorize and treat conditions according to the distinctions "hot" or "cold," maintaining the balance of these distinctions is essential to their conception of human homeostasis. *Theobroma cacao* is classified as a cold remedy, primarily due to the tree's preference for understory growth and humid conditions. In Eastern Guatemala, pathologies characterized by fever and pain are typically treated with "cool" remedies [42]. *Cacao* has also been historically employed as a remedy for influenza, inflammation, and infection, which have all been supported by research on its fruit (bean). In vivo studies support *T. cacao*'s antioxidant, anticancer, and antimalarial properties [43].

Olea europaea

Olea europaea, commonly known as "the olive tree," is one of the first trees ever cultivated and has been integral to Mediterranean culture for approximately 5,000 years. This tree is praised theologically as a gift from God, and it is mentioned in both the Quran and the Bible. The olive tree is said to symbolize peace, dignity, abundance, wisdom, and health. Olive oil is produced at approximately 2,000,000 tons/year and comprises about 4% of global vegetable oil production. To this day, olive oil and table olives remain fundamental components of the Mediterranean diet, with consumption expanding on a global scale. Traditionally, olive oil has been used in preventative medical care as a cardioprotective and gastroprotective agent. Studies suggest olives may help prevent coronary disease as well as certain cancers; this effect is largely attributed to the presence of monounsaturated fatty acids and phenolic compounds like vanillic acid in the fruit. *Olea europaea* is also a proven source of natural antioxidants, like the phenolic compound hydroxytyrosol [44].

Studies have recently discovered that olives are plentiful in ECL compounds OEA), LEA, palmitoylethanolamide, oleoylphenylalanine, 2-oleoylglycerol, 2-linoleoylglycerol, 1-oleoylglycerol, and 1-linoleoylglycerol [7]. These findings suggest that the clinical benefits of olives result from a greater range of compounds than the phenolic compounds previously believed to demonstrate the most significant pharmacological effects [7]. Moreover, even though ECL compounds are known to have significant effects on the human body, more research is required to elucidate their contribution to *O. europaea*'s medicinal benefits [7].

Amorpha fruticosa L.

Amorpha fruticosa L. is a shrub native to south-eastern Canada, the United States, and northern Mexico within the plant family Fabaceae. This shrub is commonly known as desert false indigo or false indigo-bush, which comes from its traditional use as a blue dye. Traditional medicinal uses include stomach pain, intestinal worms, eczema, and rheumatism [45]. *Amorpha fruticosa* contains amorfrutins, which are bioactive compounds, all of which contain a cannabinoid backbone [5]. Amorfrutins are also present within *H. umbraculigerum* (see Figure 1) and *Glycyrrhiza foetida*, also known as licorice. Amorfrutins are natural activators of PPARγ and exert potent anti-inflammatory effects [5]. The anti-inflammatory effects of amorfrutins are associated with the inhibition of nuclear transcription factor-κB, which is a key regulator of inflammation in the body. The therapeutic potential for the treatment of diabetes has become increasingly prevalent; recent pre-clinical rodent models have shown these compounds can stimulate the metabolism of lipids and glucose and decrease insulin resistance. These effects are believed to be at least partially mediated by amorfrutins' agonism of PPARγ, though the full mechanism of action has yet to be fully elucidated. *Amorpha fruticosa* expresses amorfrutin A, B, and C, each of which are found throughout the shrub in varying quantities with particularly high levels expressed in the seeds [46].

Salvia divinorum

Salvia divinorum, a plant belonging to the family Lamiaceae, or mint family, this plant is naturally occurring in only one small region of Oaxaca, Mexico. This plant contains one of the strongest naturally occurring psychoactive compounds known to man, salvinorin A. Salvinorin A is a diterpene that demonstrates high selectivity for the kappa opioid receptor, the receptor through which it mediates its psychoactivity. It has been used traditionally for shamanic purposes as well as in traditional Aztec medicine for gastrointestinal upset. Recent research has revealed that this plant has both anti-inflammatory and analgesic effects [47]. Though salvinorin A has very low affinity for cannabinoid receptors and does not inhibit ECB degradation, research does suggest it may have an indirect cannabimimetic effect [1]. A 2009 preclinical study by Fichna et al. revealed that salvinorin A interacted with CB1/κ-opioid receptor dimers [48]. This dimerization can modulate the function of both these receptors [49].

Conclusion

Traditional medicines, like the botanical therapeutics introduced in this chapter, make a significant contribution to healthcare across the world. These practices continue to be employed in the countries in which they originated, and methods utilizing these techniques have since expanded to other terrains and populations. The rationale behind this pervasiveness ranges from maintaining a connection to cultural practices, treatment of diseases that lack modern medical treatments for malignant disease, and lack of accessibility to allopathic care. Understanding the potential of plants and other natural products is at the foundation of modern medicine. There are still gaps in our understanding of what properties the 120 known cannabinoids within *Cannabis* sp. have to offer mankind. Pursuing exploration into the potential clinical applications of natural compounds like cannabinoids opens a path for innovation and understanding, both in the realm of traditional and allopathic medicine.

Furthermore, gaining insight into the host of species that express cannabinoids outside of *Cannabis* makes one critically analyze the stigma associated with *Cannabis*. Observing the complexity of cannabinoids in their structure and mechanism of action brings these chemicals to a new light. This deconstruction allows one to explore the dynamic relationships that lead to the production of this plant's cannabinoids, the chemical constituents that give rise to its effects, psychoactive and otherwise. This intentional act can dismantle some of the taboo that surrounds *Cannabis*. It allows one to recognize that compounds found within *C. sativa* are not much different from those recognized in sources across the natural world. May it be the CBGA produced by *H. umbraculigerum* or the hundreds of sources of (E)-BCP, cannabinoids have enduring effects that echo beyond the psychoactivity associated with THC. Investigation into the therapeutic potential of lesser-known cannabinoids, both within and outside of the *Cannabis* species, supports the clinical utilization of these compounds as an anti-inflammatory, antibacterial, gastrointestinal aid, and beyond. Though research and implementation of these compounds and plant products in a clinical setting are still limited, studies like those examined within this chapter demonstrate the fields' continued growth.

References

[1]   Morales, P. et al. (2017). Molecular Targets of the Phytocannabinoids: A Complex Picture. Progress in the chemistry of organic natural products, 103, 103–131.

[2]   Hui-Chen, L., Mackie, K. (2021). Review of the endocannabinoid system. Biological Psychiatry: Cognitive Neuroscience and Neuroimaging, 6(6), 607–615.

[3]   Maccarrone, M. et al. (2015). Endocannabinoid signaling at the periphery: 50 years after THC. Trends in pharmacological sciences, 36(5), 277–296.

[4]   Sagar, K.A., Gruber, S.A. (2018). Marijuana matters: reviewing the impact of marijuana on cognition, brain structure and function, & exploring policy implications and barriers to research. International review of psychiatry, 30(3), 251–267.

[5]   Gülck, T., Møller, B.L. (2020). Phytocannabinoids: origins and biosynthesis. Trends in plant science, 25(10), 985–1004.

[6]   Pollastro, F. et al. (2017). Amorfrutin-type phytocannabinoids from Helichrysum umbraculigerum. Fitoterapia, 123, 13–17.

[7]   Kocadağlı, T., Yılmaz, C., Gökmen, V. (2024). Effects of fermentation and alkalisation on the formation of endocannabinoid-like compounds in olives. Food Chemistry, 457, 140164.

[8]   De Petrocellis, L., Di Marzo, V. (2009). An introduction to the endocannabinoid system: from the early to the latest concepts. Best practice & research Clinical endocrinology & metabolism, 23(1), 1–15.

[9]   Shijia, L. et al. (2024). Cannabigerol (CBG): A Comprehensive Review of Its Molecular Mechanisms and Therapeutic Potential. Molecules (Basel, Switzerland), 29(22), 5471. 20 Nov. doi: 10.3390/molecules29225471.

[10]   McPartland, J.M., Glass, M., Pertwee, R.G. (2007). Meta-analysis of cannabinoid ligand binding affinity and receptor distribution: interspecies differences. British journal of pharmacology, 152(5), 583–593.

[11]   Laprairie, R.B. et al. (2015). Cannabidiol is a negative allosteric modulator of the cannabinoid CB1 receptor. British journal of pharmacology, 172(20), 4790–4805.

[12]   Gertsch, J., Pertwee, R.G., Di Marzo, V. (2010). Phytocannabinoids beyond the Cannabis plant–do they exist? British journal of pharmacology, 160(3), 523–529.

[13]   Gonçalves, E.C.D. et al. (2020). Terpenoids, cannabimimetic ligands, beyond the cannabis plant. Molecules, 25(7), 1567.

[14]   Erman, P. et al. (2023). Parallel evolution of cannabinoid biosynthesis. Nature plants, 9(5), 817–831.

[15] Lourens, A.C.U., Viljoen, A.M., Van Heerden, F.R. (2008). South African Helichrysum species: A review of the traditional uses, biological activity and phytochemistry. Journal of Ethnopharmacology, 119(3), 630–652.

[16] Burgel, L. et al. (2020). Impact of growth stage and biomass fractions on cannabinoid content and yield of different hemp (Cannabis sativa L.) genotypes. Agronomy, 10(3), 372.

[17] Tahir, M.N. et al. (2021). The biosynthesis of the cannabinoids. Journal of cannabis research, 3(1), 7.

[18] Bakali, M., Tshisikhawe, M.P., Ligavha-Mbelengwa, M.H. (2015). Ethnobotanical survey of Androstachys johnsonii Prainin Matshena village, Mutale local municipality. South African Journal of Botany, 98, 206.

[19] Nyazema, N.Z., Chanyandura, J.T., Egan, B. (2024). The use and potential abuse of psychoactive plants in southern Africa: an overview of evidence and future potential. Frontiers in pharmacology, 15, 1269247.

[20] Maffei, M.E. (2020). Plant natural sources of the endocannabinoid (E)-β-caryophyllene: A systematic quantitative analysis of published literature. International journal of molecular sciences, 21(18), 6540.

[21] Francomano, F. et al. (2019). β-Caryophyllene: a sesquiterpene with countless biological properties. Applied Sciences, 9(24), 5420.

[22] De oliveira santos, M. et al. (2022). Copaifera langsdorffii Desf.: A chemical and pharmacological review. Biocatalysis and Agricultural Biotechnology, 39, 102262.

[23] Takooree, H. et al. (2019). A systematic review on black pepper (Piper nigrum L.): from folk uses to pharmacological applications. Critical reviews in food science and nutrition, 59(sup1), S210–S243.

[24] Pandey, A. et al. (2014). Synergistic study of antioxidant potential of different spices and their bioactive constituents. International Journal of Pharmaceutical Sciences and Research, 5(8), 3267.

[25] Boonrueng, P. et al. (2022). Combination of curcumin and piperine synergistically improves pain-like behaviors in mouse models of pain with no potential CNS side effects. Chinese Medicine, 17(1), 119.

[26] Khairani, S. et al. (2021). The Potential use of a Curcumin-Piperine Combination as an Antimalarial Agent: A Systematic Review. Journal of tropical medicine, 2021(1), 9135617.

[27] Cho, H.-K., Park, C.-G., Lim, H.-B. (2024). Construction of a Synergy Combination Model for Turmeric (Curcuma longa L.) and Black Pepper (Piper nigrum L.) Extracts: Enhanced Anticancer Activity against A549 and NCI-H292 Human Lung Cancer Cells. Current Issues in Molecular Biology, 46(6), 5551–5560.

[28] Cock, I.E., Cheesman, M.A.T.T.H.E.W. (2018). Plants of the genus Syzygium (Myrtaceae): A review on ethnobotany, medicinal properties and phytochemistry. Bioactive compounds of medicinal plants: Properties and potential for human health, 35–84.

[29] Sharifi-Rad, M. et al. (2018). Echinacea plants as antioxidant and antibacterial agents: From traditional medicine to biotechnological applications. Phytotherapy Research, 32(9), 1653–1663.

[30] Oláh, A. et al. (2017). Echinacea purpurea-derived alkylamides exhibit potent anti-inflammatory effects and alleviate clinical symptoms of atopic eczema. Journal of dermatological science, 88(1), 67–77.

[31] Tucci, P. et al. (2023). The Plant Derived 3-3′-Diindolylmethane (DIM) Behaves as CB2 Receptor Agonist in Prostate Cancer Cellular Models. International Journal of Molecular Sciences, 24(4), 3620.

[32] Asgarpanah, J., Khoshkam, R. (2012). Phytochemistry and pharmacological properties of Ruta graveolens L. Journal of Medicinal Plants Research, 6(23), 3942–3949.

[33] Wilson, G. et al. (2023). Exploring the therapeutic potential of natural compounds modulating the endocannabinoid system in various diseases and disorders. Pharmacological Reports, 75(6), 1410–1444.

[34] Nazir, M. et al. (2021). Meroterpenoids: A comprehensive update insight on structural diversity and biology. Biomolecules, 11(7), 957.

[35] Hussain, T. et al. (2019). Demystifying the liverwort Radula marginata, a critical review on its taxonomy, genetics, cannabinoid phytochemistry and pharmacology. Phytochemistry Reviews, 18(3), 953–965.

[36] Reis, M.H. et al. (2020). Shared binding mode of perrottetinene and tetrahydrocannabinol diastereomers inside the CB1 receptor may incentivize novel medicinal drug design: Findings from an in silico assay. ACS Chemical Neuroscience, 11(24), 4289–4300.

[37] Chicca, A. et al. (2018). Uncovering the psychoactivity of a cannabinoid from liverworts associated with a legal high. Science advances, 4(10), eaat2166.

[38] Popescu, R., Kopp, B. (2013). The genus Rhododendron: an ethnopharmacological and toxicological review. Journal of Ethnopharmacology, 147(1), 42–62.

[39] Gardea, A.A. et al. (2017). Cacao (Theobroma cacao L.). In: Fruit and Vegetable Phytochemicals: Chemistry and Human Health. 2nd Edition, pp. 921–940.

[40] Fusar-Poli, L. et al. (2022). The effect of cocoa-rich products on depression, anxiety, and mood: A systematic review and meta-analysis. Critical reviews in food science and nutrition, 62(28), 7905–7916.

[41] Dillinger, T.L. et al. (2000). Food of the gods: cure for humanity? A cultural history of the medicinal and ritual use of chocolate. The Journal of nutrition, 130(8), 2057S–2072S.

[42] Kufer, J., Grube, N., Heinrich, M. (2006). Cacao in Eastern Guatemala--a sacred tree with ecological significance. Environment, Development and Sustainability, 8(4), 597–608.

[43] Ishaq, S., Jafri, L. (2017). Biomedical importance of cocoa (Theobroma cacao): significance and potential for the maintenance of human health. Matrix Science Pharma, 1(1), 1–5.

[44] Uylaşer, V., Yildiz, G. (2014). The historical development and nutritional importance of olive and olive oil constituted an important part of the Mediterranean diet. Critical Reviews in Food Science and Nutrition, 54(8), 1092–1101.

[45] Kozuharova, E. et al. (2017). Amorpha fruticosa–A noxious invasive alien Plant in Europe or a medicinal plant against metabolic disease? Frontiers in Pharmacology, 8, 333.

[46] Simeonova, R. et al. (2022). A study on the safety and effects of Amorpha fruticosa fruit extract on spontaneously hypertensive rats with induced type 2 diabetes. Current Issues in Molecular Biology, 44(6), 2583–2592.

[47] Schnarr, B. (13 Jun. 2024). "What Is Salvia?". The Psychedelic Pulse.

[48] Fichna, J. et al. (2009). Salvinorin A inhibits colonic transit and neurogenic ion transport in mice by activating κ-opioid and cannabinoid receptors. Neurogastroenterology & Motility, 21(12), 1326–e128.

[49] Russo, E.B. (2016). Beyond cannabis: plants and the endocannabinoid system. Trends in pharmacological sciences, 37(7), 594–605.

Jahan Marcu

Who owns a forbidden plant? Patents, power, and the architecture of exclusion

From prohibition to protection

You might reasonably ask: why is a PhD writing about intellectual property (IP), a subject usually left to lawyers and patent examiners? The answer is simple. After more than two decades in the *Cannabis* world, I've been pulled into legal battles of nearly every stripe. My role is almost always that of expert witness – whether in civil or criminal skirmishes in state courts or in amicus briefs that found their way to the Supreme Court. And if you sit in enough courtrooms, you begin to notice that the disputes are rarely about science in the abstract; they are about ownership.

One case stands out. I was retained to assess genetic testing in a dispute between two former partners, one of whom styled himself an "OG grower." His claim was that his partner was cultivating his plants and hybrids thereof, and that he could identify them by sight alone. If they would only let him inside the warehouse the issue could be settled promptly: wherever his finger landed, that plant was his property. OG Grower alleged that he had developed 20 unique varieties, though he had never once submitted a sample for potency or genetic analysis. Imagine, then, the seriousness of a multimillion-dollar enterprise, involving an IP dispute, hinging on the pointing finger of a grower.[1]

Despite the dancing eyes and bubbly bon vivant attitudes associated with the business of *Cannabis*, the industry can be quite taciturn when it comes to certain legal issues. For most of the twentieth century, the words *Cannabis* and *IP* would have seemed mutually exclusive. *Cannabis* was contraband; ideas about its cultivation or use belonged to an underground economy, not to patent offices or trademark registries. Yet today, the industry finds itself in an inversion: the very plant once deemed illicit is now the focus of sophisticated and solipsistic strategies to protect innovation, secure brands, and attract investment.

At the heart of this transformation is the concept of *property* itself. In law, the most fundamental incident of property is not use or enjoyment, but the *right to exclude* [1]. A farmer "owns" land because she can exclude trespassers. IP extends this right to the intangible: ideas, inventions, creative expressions, and distinctive signs [2, 3]. Unlike land, ideas are not scarce; they can be replicated endlessly. IP law therefore exists not to ration a finite resource, but to solve a different economic problem – the *free rider*

---

1 For those of you interested about the outcome, with no baseline for comparison the DNA tests were inconclusive. In the eyes of the court, OG Grower could prove that any plants in the warehouse were his intellectual property.

dilemma [4]. Without protection, a competitor can copy a costly innovation at marginal expense, outcompeting the originator and chilling future investment. IP law intervenes to align private incentives with social benefit: temporary monopolies in exchange for disclosure and progress. And the inverse is true as well, IP can be an enemy of innovation, keeping out competitors that might improve on a technology.

Cannabis – and now psychedelics – highlight this dynamic in stark relief. For decades, prohibition suppressed formal innovation and kept knowledge in oral traditions, underground networks, or unpublished lab notebooks. The result is a patchy "prior art" record. Prior art is the body of publicly available knowledge used to judge whether an invention is novel and nonobvious. In patent law, prior art can include earlier patents, published applications, journal articles, or any disclosure accessible to the public. In the absence of indexed journals or patent filings, examiners risk granting overly broad or weak patents. Conversely, innovators today must navigate whether their discoveries are truly novel, or whether they echo knowledge held by communities who never had the chance to publish or protect it.

The myth that "you can't patent a plant" persists in public discourse, but it is only partly true [5]. While naturally occurring plants cannot be patented merely for existing in nature – a principle established in cases like *Funk Brothers* (1948) – human modifications can be [6]. In *Diamond v. Chakrabarty*, the U.S. Supreme Court confirmed that engineered organisms fall within patentable subject matter [7]. Hops (*Humulus Lupulus* L.), the *Cannabis* plant's closest living relative, has its own lineage of patented varieties in beer brewing – each the product of nearly a decade's labor to show that what you've bred is not merely different, but distinct enough to deserve a name and a fence around it [8]. That line of reasoning has since opened the door for genetically modified or specially cultivated *Cannabis*, novel extraction processes, and engineered formulations to receive patent protection. Indeed, US agencies themselves have secured *Cannabis*-related patents, including claims on cannabinoids as neuroprotectants [9].

From this perspective, IP serves as a bridge from prohibition to protection. It provides companies with a legal scaffold to differentiate themselves in a marketplace where traditional financing, federal legality, and cross-border trade remain uncertain. By converting intangible ideas into property, IP regimes allow *Cannabis* and psychedelics companies to attract capital, secure licensing deals, and structure partnerships – even in a fragmented regulatory environment. This scaffolding, however, comes with caution: if the past century of prohibition was about exclusion from legality, the coming decades must guard against exclusion through monopolization. The OG grower's finger may have been an eccentric claim to property, but it reminds us of the essential truth that animates this chapter: ownership, in *Cannabis* and psychedelics alike, still pivots on who is empowered to point and say, "this one is mine."

Table 1: Cannabis patents index.

Number of *Cannabis*-related patents: 8,719
Of those that are for ligands, drug compositions, and combinations: 3,366
Number of patents for cultivation: 800
Most common type of *Cannabis* processing patents: extraction
Least common type of *Cannabis* processing patents: preservation
Number of *Cannabis* patents for oral delivery: 747
Minimum number of *Cannabis* coffee-related patents: 10
Number of non-oral delivery patents: 425
Minimum number of those that are for delivery with adult toys: 1
Number of patents for detecting *Cannabis* products in people: 323
Country with the most *Cannabis* patents: the United States
Country with second most: the United Kingdom
Country with fifth most: Israel
Country with 13th most: China
Number of companies holding over 100 *Cannabis* patents: 5
Number of patents held by BASF, a German company: 140
Number of patents held by GW Pharma, a UK company: 132
Number of patents held by Nicoventures Trading, a UK tobacco company: 129
Number of patents held by Sanofi-Aventis, a French company: 117
Number of patents held by Bayer, a German company: 114
Number held by the US company with the most *Cannabis*-related patents: 59
Number of countries with assignees holding three or more patents: 30
Number countries with entities holding 100 or more US cannabis patents: 8
Number of patents describing home delivery systems for *Cannabis*: 6
Of those that involve drone delivery: 2

Data derived from Cannabis Patents Quantaa Blog by Ruth Fisher, PhD (October 8, 2025).

The legal patchwork (and why IP still matters)

The *Cannabis* and psychedelic industries operate within a patchwork of laws that can shift drastically from one jurisdiction to the next. Nowhere is this more evident than in the interplay between prohibition and protection: while companies may be barred from commercializing *Cannabis* in one jurisdiction, they can still secure IP rights in another (see Table 1 for index of patents).

At the US federal level, *Cannabis* remains a Schedule I controlled substance under the Controlled Substances Act, while hemp – *Cannabis* with less than 0.3% tetrahydrocannabinol (THC) – was legalized under the 2018 Farm Bill [10]. Psychedelics such as psilocybin and 3,4-methylenedioxymethamphetamine (MDMA) remain federally prohibited, though the Food and Drug Administration (FDA) has granted "breakthrough therapy" designations to psilocybin and MDMA-assisted therapies [11]. These substances thus occupy a paradoxical position: still federally prohibited, yet accorded a form of protective status within certain pathways (see Chapter 8 – "Pandora's plant: a series of entirely predictable absurdities").

Despite this illegality, patents remain available. Courts have consistently held that subject matter is not unpatentable merely because it involves controlled substances. *Diamond v. Chakrabarty* (1980) affirmed that engineered life forms are patentable, distinguishing them from natural products in *Funk Brothers Seed Co. v. Kalo Inoculant Co.* (1948). As a result, companies have obtained US patents covering *Cannabis* extracts, formulations, and cultivation techniques (see Table 1). Federal trademark protection remains barred under the Lanham Act's "unlawful use" doctrine, which prevents marks associated with federally illegal goods from being registered, even as state and common-law trademark protections have developed in parallel [12]. In the United States, you can patent the process for making *Cannabis* extract, but you cannot trademark the name that tells anyone what it is. It's as if Coca-Cola could have patented a recipe with real cocaine, but was barred from protecting the Coca-Cola name – while a soda with no cocaine at all could stroll in and trademark "Cocaine." See Table 2 for more insight.

Table 2: Sidebar: patents, trademarks, and copyright – the psychedelics and cannabinoid problem.

IP law doesn't treat all protections the same way:
- **Patents** ask: *Is it new, useful, and nonobvious?*
- **Trademarks** ask: *Is it used lawfully in commerce, and does it tell people who made it?*
- **Copyright** asks: *Did you fix an original creative work in some tangible form?*

Put those together in cannabis and psychedelics, and you get some curious results:
- **Patents**: You could patent a novel extraction method for psilocybin or a unique cannabinoid formulation, even though the underlying substances are Schedule I.
- **Trademarks**: You couldn't federally register a mark for "Psilocybin Gummies" or "THC soda" because the products remain unlawful under federal law.
- **Copyright**: You *could* copyright the psychedelic poster advertising those same gummies or a trippy jingle about THC soda – even if selling them would be illegal.
- **The twist**: A company could federally trademark the name "Psilocybin" for a computer app or the word "Marijuana" for a clothing line – so long as the goods have nothing to do with the drug itself.

In the IP world of controlled substances, your recipe may be patentable, your song about it copyrightable, but your brand name unprotectable – unless it describes something else entirely. The result is a landscape where science can be protected, but branding cannot. Or, put it simply: "Your secret sauce is protectable – but not the name on the bottle."

IP outside the United States

In Europe, *Cannabis*-related patents can be pursued through the European Patent Office (EPO), provided they meet the usual standards of novelty, inventive step, and industrial applicability. But trademarks, however, follow a different path. The European Union Intellectual Property Office (EUIPO) runs the EU trademark system, and it applies an extra filter. EUIPO may reject applications that violate "public policy or accepted principles of morality" [13]. On paper, this sounds principled; in practice, it depends on the individual morals of the examiner reviewing your paperwork. Are they a devout Catholic or a relaxed secularist? Did they skip lunch and arrive at your file in need of coffee, primed to take offense at the smallest provocation? The point is that the outcomes have been wildly inconsistent and we can only imagine the reasons. Some *Cannabis* accessories and hemp-derived products have been registered, while marijuana-specific marks are often denied. Individual member states add further complexity: Germany has recognized medical *Cannabis* marks, while others, such as France, have been stricter. Put simply: the EPO may reward you for your clever process, while the EUIPO frowns at the name you hoped to put on the label.

Canada legalized recreational *Cannabis* in 2018, creating the world's first G7-scale federally legal *Cannabis* market. But this celebration came with a gag order. Their *Cannabis* Act imposes some of the harshest restrictions on branding and marketing [14]. Trademarks can be registered with the Canadian Intellectual Property Office, but packaging, advertising, and promotional rules strip away much of their expressive power. Plain packaging requirements, bans on lifestyle branding, and severe restrictions on logos and colors limit how far a registered mark can travel in the marketplace. In practice, the packaging of Canadian *Cannabis* often arrives dressed like a stage extra – plain, uniform, and deliberately drained of character – less consumer product than placeholder, a blank carton that might as well read *Generic Vice, Contents Variable.*

Sure, a company can invest in responsible, eye-catching marketing and product infographics but a consumer in Canada will never see them. Which raises the question: what is the point of registering mute (trade)marks? In effect, Canada demonstrates how legality does not necessarily translate into IP freedom, and in some ways makes trademarks whisper when they were meant to shout.

In the Global South, debates around *Cannabis* and IP often intersect with postcolonial questions of knowledge and sovereignty. South Africa has moved toward medical *Cannabis* legalization but faces tension between patenting new cultivars and respecting Indigenous landraces long used by local communities [15]. For example, communities that long tended "Durban Poison" in South Africa may find their hard work transformed into claims by outside actors who never tilled the soil or joined a ceremony. The challenge is whether new patents on "improved" cultivars risk enclosing what local communities long treated as shared resources.

Brazil, by contrast, has historically excluded plant varieties from patentability under its industrial property law, channeling protection instead through plant variety rights systems [16]. This approach reflects both a policy choice and a cultural one: a reluctance to collapse agricultural biodiversity into private monopolies. The Global South reveals the limitations of conventional IP tools when they collide with local cultural and legal frameworks. This feels like the layering a foreign system on top of existing knowledge traditions. Even with the best of intentions, the irony is hard to miss: innovation is rewarded not for centuries of stewardship, but for writing it down in the right legal format.

Why IP still matters in the patchwork

Across these jurisdictions, IP emerges as one of the few stable tools in an otherwise unstable regulatory environment. In the United States, patents and trade secrets provide leverage where banking and tax regimes constrain growth (notably IRS § 280E, which denies standard deductions for businesses dealing in controlled substances without federal licensure). In Europe, patents can be globalized even where trademarks stumble on morality clauses. In Canada, companies still rely on IP portfolios to raise capital and secure international licensing, despite branding restrictions at home.

For psychedelics, where pharmaceutical pathways dominate, patents and regulatory exclusivities under acts like Hatch-Waxman in the United States or European Medicines Agency data protection rules in Europe are not just protective – they are existential [17]. The billion-dollar cost of clinical development cannot be recouped without enforceable exclusivity [18].

Thus, while the legal patchwork complicates commerce, it amplifies the strategic importance of IP. Patents, trademarks, and exclusivities are not just tools of competition; they are anchors of legitimacy and value in industries defined by prohibition, stigma, and fragmentation.

The IP toolbox: patents, trademarks, trade secrets, and beyond

Cannabis and psychedelics companies must navigate a hybrid IP system – formal legal doctrines that establish eligibility and enforcement on one side, and a patchwork of scholarly and practitioner insights that reveal the limits and opportunities on the other. Together, they illustrate both the ingenuity and fragility of IP in this space.

The U.S. Supreme Court long ago drew the boundary: "laws of nature, physical phenomena, and abstract ideas are not patentable" (*Funk Bros. Seed Co. v. Kalo Inoculant Co.*, 333U.S. 127, 130 (1948)), yet "anything under the sun that is made by man" can be (*Diamond v. Chakrabarty*, 447U.S. 303, 309 (1980)). This framework most likely excludes wild-type *Cannabis* but admits there is a pathway for engineered strains, extraction methods, and dosage regimens.

Scholarly commentary situates this in practice. Let us contrast the United States and EU briefly: In the United States, trademarks and patents are difficult due to strict federal illegality. For plant varieties (aka cultivars or strains), the Plant Variety Protection Act (PVPA) covers sexually reproduced plants and explicitly requires confirmation that the variety has less than 0.3% THC. In the EU, as mentioned above for trademarks, it depends on subjectively assessing "morality," which complicates the granting the rights among 27 member states. The EU is also generally considered to be much stricter about their *novelty* standard [19].

So, before you get worried about "big pharma" or "Multi-State Operator (MSOs)," patenting your grandpa's Silver Haze lineage, we should further underscore the technical challenges of patenting a plant via the scarcity of *Cannabis* plant patents; fewer than 50 utility patents exist for cultivars and genetics, though the website Leafly lists more than 8,000 strains (aka varieties) [20]. Patented *Cannabis* products would certainly be perceived as higher quality or more desirable – exclusivity not only enhances market presence but also supports premium pricing strategies [21]. Once you've demonstrated you have created a novel cultivar – say, one that reliably smells of ammonia with sour notes and sold under the name *Cheetah Piss*[2] – the next step might be trademarking that name. After all, you wouldn't want a bunch of sub-par, knockoff *Cheetah Piss* undermining your reputation.

As mentioned earlier, the Lanham Act remains the foundation of US trademark law,[3] but the *unlawful-use doctrine* prevents federal registration for *Cannabis* goods, leading to inconsistent judicial outcomes.[4] You could cultivate it, even patent a novel version of it, but, sorry, you probably will not be able to federally register *Cheetah Piss* as a trademark for *Cannabis* flower. Practitioners have identified workarounds, a term coined "trademark laundering" to describe registering marks for ancillary goods (e.g., t-shirts or vaporizers) and informally extending them to *Cannabis* [22]. That means a grower might protect the cultivar's name as a logo on a hoodie or as a sports drink (*if it contained no Cannabis at all*). The irony is that the more outrageous the name, the easier it may be to register in an unrelated category, while the actual *Cannabis* goods carrying that name remain unprotectable at the federal level. If you're struggling with protecting your illicit IP, take respite in knowing that patent struggles have a long history in the United States, Thomas Jefferson himself once petitioned Congress for trademark protection, a request ignored until nearly a century later [23]. Outside the United States, Canada's contradictory approach suggests marks are registrable, but plain-packaging provisions under the *Cannabis* Act render them mute; a registered mark exists, but cannot speak [24].

---

2 "Cheetah Piss is funky just like Cat Piss, with skunky, ammonia, and sour notes, reminding consumers of those strains with unique qualities." See https://www.leafly.com/strains/cheetah-piss.

3 Lanham Act, ch. 540, 60 Stat. 427 (1946) (codified as amended at 15 U.S.C. §§ 1051–1141n).

4 See, e.g., *In re Brown*, 119 U.S.P.Q.2d 1350 (T.T.A.B. 2016) (rejecting cannabis-related marks under unlawful-use doctrine).

Trade secrets: the silent backbone

Trade secrets cover cultivation methods, nutrient mixes, extraction techniques, and customer lists, so long as reasonable secrecy measures are used (nondisclosure agreements (NDAs)), access controls, need-to-know compartmentalization). The Defend Trade Secrets Act of 2016 extends federal jurisdiction and remedies.[5]

It cannot be stressed enough how secrecy is often bolstered through contractual webs: *Cannabis*-related businesses routinely "license" IP rights; often vaguely defined, sometimes no more than trade secrets, brand names, or know-how – and extend them across state lines despite legality gaps. This practice underscores the practical reality: patents disclose and expire, while trade secrets persist so long as the secret holds. A smart portfolio often blends the two – a patent staking out the broad inventive claim, while the unspoken details (the optimal grow room humidity curve, the proprietary filter design) remain locked in the vault. But as in any industry, these agreements are only as strong as your ability – and financial willingness – to enforce them. An NDA without the resources to litigate is less a fortress wall than a polite "Keep Out" sign.

Running alongside patents and trade secrets are regulatory exclusivities, sometimes called "shadow patents." In the United States, Hatch-Waxman grants 5 years of data exclusivity for new chemical entities, 3 years for new clinical studies, and 30-month stays if a listed patent is challenged.[6] In Europe, the EMA provides the "8 + 2(+1)" model: 8 years of data exclusivity, 2 years of market protection, and a bonus year for new indications [25].

GW Pharmaceuticals' litigation over Epidiolex® illustrates the punch these protections can carry. By asserting patents on cannabidiol delayed generic entry by 30 months, generating hundreds of millions in additional revenue [26]. Firms with established IP portfolios can use these mechanisms not only to defend market share but also to shape the pace of regulatory change itself [21]. This means the first mover's IP strategy doesn't just block competitors; it influences how the regulator defines "safe" and "effective" in that therapeutic area. Exclusivity doesn't just protect your product; it tells the regulator how fast the rest of the field is allowed to run. In *Cannabis* and psychedelics, where first movers don't just win the race – they set the length of the track.

Each tool is partial, and each faces structural hurdles. Patents are hemmed in by novelty requirements and a thin record of prior art; trademarks are clipped by unlawful-use doctrines, plain-packaging rules, or "morality" assessments; trade secrets can slip away through reverse engineering or a disgruntled employee; exclusivities, by design, expire. Yet, when deployed together, they form a moat – uneven, leaky, but still wide enough to slow competitors and reassure investors.

---

5 Defend Trade Secrets Act of 2016, 18 U.S.C. § 1836.

6 Drug Price Competition and Patent Term Restoration Act of 1984 (Hatch-Waxman Act), Pub. L. No. 98–417, 98 Stat. 1585.

The harder question is not whether these tools work, but when to use them. Should you disclose or conceal? Announce your invention to the world or lock it in a vault? History suggests that civilizations have been undone both by secrets no one shared and by revelations no one controlled. The same is true here: in *Cannabis* and psychedelics, the real drama begins when you ask, not "Can this be protected?" but "Should this be a secret?"

And that is where the next challenge lies.

"Are you sure this should be a secret?"

In *Cannabis* and psychedelics, where much innovation grows out of informal knowledge and trial-and-error, companies often assume secrecy is safest. But secrecy is a double-edged sword: it preserves control only so long as the secret is kept, and it can leave businesses unprotected if a competitor independently rediscovers or reverse-engineers the same process. And here the lesson bites: secrets work until the competition figures out your trick. Once they do, an inventor could be left empty-handed, wishing they had written it down and claimed it first. The central strategic question, then, is deceptively simple: Are you sure this should be a secret?

What belongs in the light

Patents are disclosure-driven. Under 35 U.S.C. § 112, an inventor must describe the invention in "full, clear, concise, and exact terms" sufficient for others to reproduce it. A patent can be used to cover broader aspects of the invention while trade secret protection extends to the details, such as an optimal configuration. In exchange, a 20-year right to exclude is granted. This bargain is often worth making when:
- The innovation is **readily reverse-engineered** (e.g., chemical formulations and dosage forms).
- Financing depends on **tangible IP assets** (patents anchor valuation and licensing).
- The innovation has **broad applicability** (extraction processes, cultivation techniques).

It is true that patents promise 20 years of exclusivity, but that term is misleading. Much of it is consumed by development, trials, and regulatory review. The effective commercial life of a pharmaceutical patent is often a decade or less. Yet that decade exists only because the early disclosure staked a claim. Without it, competitors may freely replicate or even block you with their own filings.

Sometimes disclosure functions defensively: putting knowledge into the record to keep others from claiming it. In psychedelics, prior art projects like Porta Sophia[7] have deliberately published historical uses to prevent overly broad patents from enclosing common practices. In *Cannabis*, patenting extraction methods has both anchored licensing revenue and prevented later attempts to privatize techniques already widespread in the industry.

Patents, then, are not simply legal shields. They are signals, maps, and sometimes barricades – imperfect in duration, but powerful in the clarity they impose.

What belongs in the shadows

Trade secrets, by contrast, protect information that "derives independent economic value from not being generally known" if "reasonable measures" are taken. Secrecy carries its own power. Unlike patents, it has no built-in expiration; a formula or process can, in theory, endure indefinitely. The Coca-Cola recipe has lasted more than a century not because it was patented, but because it was kept in a vault. In *Cannabis*, growers guard cultivation playbooks, strain-selection heuristics, and nutrient recipes with similar zeal, passing them hand-to-hand under NDA. In psychedelics, proprietary protocols for manufacturing GMP-grade psilocybin or stabilizing a novel chemical entity are often held as trade secrets precisely because publishing them would invite imitation before a single dose reached the clinic. Secretive tactics are best reserved for:

- **Tacit knowledge**: Know-how that is difficult to document
- **Fine-tuned parameters**: Temperature curves, nutrient recipes, and proprietary microbiomes
- **Business intelligence**: Supplier/customer lists and pricing strategies

As Kamin and Moffat [22] observe, secrecy in *Cannabis* is often maintained not only by NDAs but also by elaborate "contractual webs" that purport to license IP across states, even where legality is unsettled. Sandell [27] suggests layering protection: breeders might patent plant traits or extraction methods while holding cultivation protocols as trade secrets. Cantore [21] adds that this dual approach enhances both exclusivity and perceived product quality, supporting "premium pricing strategies" even in competitive markets. As Cantore [21] implies, such a strategic decision is not binary. *Cannabis* innovators should think in terms of portfolios, where patents, trade secrets, and even trademarks operate in tandem rather than in isolation.

Trade secrets, then, are less monuments than diaries: valuable so long as they stay closed, but instantly diminished once exposed. In industries born of prohibition

---

7 Porta Sophia is a simple search tool for innovators and patent reviewers to find relevant prior art in the field of psychedelics. https://www.portasophia.org.

and informality, where much innovation has always lived off the record, shadows are as necessary as light. The art is knowing which knowledge strengthens you when spoken aloud, and which must remain written in invisible ink (see Table 3).

Table 3: The four-question test.

Before defaulting to secrecy, innovators should ask:
1. **Can it be reverse-engineered?** If yes, patents are safer than secrecy.
2. **Will disclosure strengthen your position?** Patents can attract investors, partners, and licensees.
3. **Is the value in tacit know-how or process nuance?** If so, secrecy may better preserve advantage.
4. **What happens when the secret leaks?** Patents survive publication; trade secrets die with disclosure.

Applied monetization licensing, cross-licensing and enforcement

For *Cannabis* and psychedelics companies, IP only creates value when it is actively deployed. This often takes the form of licensing agreements, cross-licenses, or enforcement actions (see Table 4). Each requires careful structuring in an industry where federal illegality, patchwork regulation, and stigma complicate traditional IP playbooks.

Table 4: Practical takeaways for licensing.

---

1. **Licensing**: Vital for monetization, especially where direct sales face legal hurdles
2. **Cross-licensing**: Anticipate patent thickets; use negotiation to avoid stalemates
3. **Enforcement**: Federal illegality complicates US actions, but patents and trade secrets remain enforceable in principle. Other types of contracts and agreements such as nondisclosures could be enforceable
4. **Global lens**: Licensing is increasingly tied to sovereignty (Jamaica), public health (Canada), or morality doctrines (EU).

Keep in mind that IP rights are not static shields or trophies, but more like dynamic tools or instruments, which are leveraged through contracts, alliances, and, when necessary, targeted actions in court systems.

Label and refer to in text.

Licensing transforms IP from a defensive right to a revenue stream. A patent, trademark, or trade secret can be licensed on an exclusive or nonexclusive basis, with royalties tied to sales, geography, or fields of use. Patents are particularly attractive in pharmaceuticals. As the Hatch-Waxman regime illustrates, patents can be listed in the FDA's Orange Book, triggering automatic 30-month stays when challenged.[8] For psy-

---

8 Ibid.

chedelics, where R&D costs are measured in the hundreds of millions, these stays and royalty streams are central to investment recovery. Licensing agreements not only provide access to cutting-edge technologies but also establish collaborative frameworks that can drive further innovation.

In fragmented industries, cross-licensing helps resolve stand-offs where multiple patents cover overlapping fields. This is already familiar in pharmaceuticals and agricultural biotechnology. As *Cannabis* patents proliferate, the same dynamics emerge. Sandell [27] notes that utility patents on *Cannabis* varieties or traits may overlap with plant patents or PVPA certificates, creating a thicket that can be resolved only by licensing, and the courts as a last resort. As is similarly observed in over industries, companies use broad method claims to create *blocking positions*, forcing competitors into negotiation.

Enforcement and the illegality hurdle

Enforcement remains the most fraught dimension. Patent suits are filed in federal courts, which must apply federal law even as the underlying product (marijuana) remains a Schedule I controlled substance. Courts have not yet squarely refused to enforce *Cannabis* patents, but uncertainty seems to have a chilling effect on litigation [28]. Trademarks face sharper limits. Federal registration is unavailable for illicit drugs under the unlawful-use doctrine. Yet, state registrations and common-law rights have been enforced, particularly in unfair competition and passing-off cases.[9]

To stress the irony: *Cannabis*-related businesses regularly assert IP rights they may not validly hold, and courts have sometimes indulged the fiction to prevent obvious unfairness [22]. It's a bit like fighting over who owns the bootleg t-shirt booth outside the stadium – everyone knows the business is unlawful, yet courts referee the squabble just to keep the knives from coming out. And in my experience, the victor is usually whoever kept better records, might win on simple principles – or, in this analogy, whoever *pressed* the cleaner copy of merchandise.

Psychedelics ≠ *Cannabis*: pharma rules, different stakes

Although *Cannabis* and psychedelics often travel together in policy debates over legalization and IP, their innovation ecosystems diverge sharply. *Cannabis* operates largely as an agricultural commodity and consumer good (vapes, bath bombs, and salves), while psychedelics (psilocybin, Lysergic acid diethylamide [LSD], and MDMA) are increasingly

---

9 Ibid.

structured as pharmaceutical products, subject to clinical trials, FDA pathways, and international drug control treaties. *Cannabis* plays a breadth game (many tools, shallow moats, and speed); psychedelics play a depth game (few tools, deep moats, and slow).

For psychedelics, the pharmaceutical model dominates. Utility patents covering molecules, formulations, and methods of treatment are the cornerstone of protection. As one analysis notes, patent rights take on heightened importance because the cost of bringing a psychedelic drug through the FDA process can approach $1 billion, inclusive of failed trials [18]. Regulatory exclusivities operate alongside patents to secure return on investment. By contrast, most *Cannabis* businesses do not seem to enter FDA pipelines, and their IP strategy relies more on plant patents, trade secrets (non-disclosures), trademarks, contracts, and state-level licensing [10].

Both *Cannabis* and psychedelics share a history of criminalization that suppressed publication, thus usually a dearth of searchable prior art. It is generally accepted that much knowledge of illicit product techniques or trade secrets, exists in underground or informal literature, not collated or indexed in a traditional manner, raising the danger that the United States Patent and Trademark Office (USPTO) may issue overly broad or even invalid patents. This has already generated controversy in psychedelics IP, where patents on known molecules or generic methods of therapy have been criticized as attempts to privatize preexisting knowledge.

The absurdity writes itself: one might as well argue that because Vikings took mind-altering substances before battle, their descendants, therefore, now own all claims to performance-enhancing psychedelics. Or perhaps, someone brewed hemp tea in a monastery in the twelfth century; thus, every CBD latte sold in Brooklyn is infringing prior art. Thin prior art leaves examiners vulnerable to precisely this kind of overreach, dressed up in legal language. See Table 5 for examples of prior art.

Enforcement and illegality doctrine

Here, too, psychedelics differ. *Cannabis* enforcement cases already test whether federal courts will honor IP rights tied to a federally illegal product. District courts have allowed trade secret and patent claims to proceed, but appellate courts have yet to decide the issue. For psychedelics, illegality questions may matter less, since most enforceable IP arises in the context of FDA-approved R&D. In these cases, patents and exclusivities are aligned with federal regulatory policy, reducing the risk that courts will refuse enforcement.

The business stakes are also different. *Cannabis* patents can enhance consumer brand power, but they exist in a diffuse market with thousands of cultivars and products. Psychedelic patents, by contrast, are few but tied to billion-dollar pharmaceutical prospects [29]. The incentive structures thus diverge: *Cannabis* firms focus on breadth (protecting strains, logos, processes), while psychedelic firms pursue depth (exclusive rights to a clinical formulation or therapeutic protocol). There are no FDA-approved

Cannabis products being sold in dispensaries, but they *are currently being sold*. Psychedelics are reaching for FDA approval, which offers breathtaking market advantages, but no revenue from sales until regulatory approval.

Regulatory outcomes are not guaranteed. In 2024, the FDA declined to approve MDMA-assisted therapy for posttraumatic stress disorder; the Complete Response Letter outlining deficiencies was made public in 2025, and reportage summarizes the agency's concerns (durability, safety, bias, trial design) [30]. These events don't foreclose future approvals, but they illustrate the stakes and timelines of the pharma route.

Table 5: Real-world examples of thin prior art and overbreadth risks.

1. U.S. Patent No. 6630507 (2003)
   - **Assignee:** U.S. Department of Health and Human Services
   - **Claim:** Cannabinoids as antioxidants and neuroprotectants
   - **Why it's seen as overbroad:** It covers *all* cannabinoids for oxidative and neuroprotective conditions, despite centuries of anecdotal and folk use for exactly those purposes. Critics often cite this as a case of the US government patenting therapeutic uses of a substance it simultaneously classified as having "no medical use."
2. U.S. Patent Application US20140322127A1 (2014)
   - **Applicant:** GW Pharmaceuticals
   - **Claim:** Use of cannabinoids to treat glioma (a type of brain cancer)
   - **Why it's controversial:** Critics argue the claims were sweeping, encompassing almost any cannabinoid combination for gliomas – conditions where some preclinical folk and scientific literature already existed
3. U.S. Patent No. 9095554 (2015)
   - **Assignee:** Biotech Institute LLC
   - **Claim:** Varieties of cannabis with "significant" levels of THC, CBD, or both
   - **Why it's seen as absurd:** Essentially attempts to cover broad swathes of existing cannabis chemotypes ("high THC," "high CBD," and "balanced"), which growers and users had recognized for decades. Often derided as an attempt to patent the obvious
4. Strain Name IP Collisions (Trademark/Patent)
   - **Gorilla Glue #4:** A strain name that triggered a lawsuit by Gorilla Glue Co. (the adhesive manufacturer). Settled in 2017; cannabis businesses had to rebrand.
   - **Girl Scout Cookies (GSC):** Another infamous example where cannabis businesses built a brand on a name directly colliding with a famous trademark. Shows how informal "prior art" (underground strain naming) collides with formal IP.

Comparative lessons and traditional knowledge: EU, Canada, and Jamaica

IP for cannabis does not evolve in a vacuum. Other jurisdictions highlight the range of doctrinal choices and policy trade-offs available to lawmakers. For example, the EU offers a unitary trademark system through EUIPO, but registration of *Cannabis-*

related marks faces public policy and morality exclusions. Member States diverge on legality, complicating the application of a unitary right. The lack of overall exclusive competence in *Cannabis* regulation means EU trademark law is applied unevenly, with morality refusals acting as a persistent barrier [31].

In contrast, Canada illustrates the other end of the spectrum. Since legalization in 2018, *Cannabis* companies have enjoyed access to the federal trademark system and can enforce marks in courts. Canadian registrations show thousands of active *Cannabis*-related trademarks, from generic descriptors to branded strain names [32]. Health Canada also engages Indigenous communities directly: the federal *Engagement with First Nations, Inuit and Métis Peoples* report emphasizes concerns around misappropriation and the need for protection of traditional knowledge in the *Cannabis* sector [33].

Jamaica presents a unique set of circumstances. In Jamaica and across the Caribbean, Rastafarian communities have long articulated *Cannabis* as a sacrament. Yet global markets frequently misappropriate Rastafari symbols, colors, and language in trademarks and branding. Jamaica is experimenting with geographical indications (GIs) for "Jamaica *Cannabis*," "Jamaica ganja," and "Jamaica hemp." GIs differ from trademarks because they function as community rights, returning economic value to producer groups in postcolonial economies. Corthésy [34] details the costs of GI registration in Jamaica (approx. JMD160,000–250,000) and the need for alignment with international standards to protect against infringement abroad. The Madrid Protocol may be a useful tool for Jamaican producers, even though domestic law still restricts the use of "Jamaica," Rastafarian symbols, and morality-laden marks [34]. Goffe [33] extends this lens to traditional cultural expressions (TCEs), arguing that Rastafari symbols and knowledge have been repeatedly misappropriated in global *Cannabis* branding, necessitating a decolonial, community-rights approach to IP.

Lessons for reform

- **EU** demonstrates the limits of morality-based refusals in a fragmented regulatory environment.
- **Canada** shows how integration into a federal system normalizes IP protection and enables Indigenous engagement.
- **Jamaica** illustrates the promise and challenges of community-based rights (GIs and TCEs), underscoring the need for equitable benefit-sharing frameworks.
- **Community-based licensing schemes** require prior informed consent and benefit-sharing with Indigenous groups.
- **Certification and collective marks** ensure authenticity in trade without full GI registration.

Together, these models suggest that US *Cannabis* IP reform cannot stop at patents and trademarks; it must also grapple with **morality doctrines, Indigenous knowledge, and community-based IP tools.**

Conclusion: choose your model, but on whose terms?

The trajectory of *Cannabis* and psychedelics IP underscores a paradox: the same plants once criminalized as contraband are now subject to some of the most sophisticated IP strategies in law. From *Funk Brothers* and *Chakrabarty* to Hatch-Waxman and EU morality clauses, the doctrine is not new. What is new is its application to industries still marked by illegality, stigma, and fragmented regulation.

Three lessons emerge from this comparative and doctrinal survey:

1. **Disclosure versus secrecy**. Innovators must continually ask, "Are you sure this should be a secret?" Patents and trade secrets are not alternatives but complements. The smartest portfolios layer disclosure and secrecy, balancing investment needs against risks of reverse engineering or misappropriation.
2. **Patchwork persistence**. The EU's morality bars, Canada's plain packaging, and Jamaica's GI experiments each illustrate how law is shaped by cultural, political, and historical contexts. No single model is exportable wholesale; US reform will require grappling with morality doctrines, federalism, and sovereignty.
3. **Equity demands**. Perhaps the sharpest challenge is ensuring that Indigenous and marginalized communities, long excluded from legal markets, are not dispossessed again – this time by patents, trademarks, or GIs. International frameworks like TRIPS and Nagoya only partially address these concerns; genuine equity will depend on domestic laws that recognize traditional knowledge, cultural expressions, and benefit-sharing rights.

For psychedelics, the stakes are especially high: clinical development costs and regulatory exclusivities put IP at the very center of viability. For *Cannabis*, IP functions more diffusely – as a mix of plant patents, trade secrets, and branding strategies – but it still underwrites legitimacy and investment in a federally ambiguous market.

Cannabis and hemp are "multidimensional goods" and no single IP tool can capture all its dimensions. This chapter suggests the same of psychedelics. The challenge is not whether *Cannabis* and psychedelics will be protected by IP – that is already underway – but whether those protections will be effective and calibrated to balance innovation, equity, and access.

The choice is not only legal but also political: who benefits when prohibition gives way to protection?

References

[1]   *Lingle v. Chevron U.S.A. Inc.*, 544 U.S. 528. (2005). Justia U.S. Supreme Court Center. https://supreme.justia.com/cases/federal/us/544/528/

[2]   Rosenblatt, E. L. (2013). Intellectual property's negative space: Beyond the utilitarian. Florida State University Law Review. 441.

[3]   Merrill, T. W. (n.d.). *Property and the right to exclude II.* Columbia Law School Scholarship Archive. (Accessed October 1, 2025) https://scholarship.law.columbia.edu/faculty_scholarship/937/

[4]   Palmer, T. G. (1990). Are patents and copyrights morally justified? The philosophy of property rights and ideal objects. Harvard Journal of Law & Public Policy, 13, pp.817–849.

[5]   *7 U.S.C. Chapter 57 – Plant Variety Protection.* (n.d.) Legal Information Institute. Accessed October 1, 2025 https://www.law.cornell.edu/uscode/text/7/chapter-57

[6]   *Funk Brothers Seed Co. v. Kalo Inoculant Co.*, 333 U.S. Vol. 127. (1948). Justia U.S. Supreme Court Center. https://supreme.justia.com/cases/federal/us/333/127/

[7]   *Diamond v. Chakrabarty*, 447 U.S. 303. (1980). Justia U.S. Supreme Court Center. https://supreme.justia.com/cases/federal/us/447/303/

[8]   Guimarães, B. P., Nascimento, P. G. B. D., & Ghesti, G. F. (2021). Intellectual property and plant variety protection: Prospective study on hop (*Humulus lupulus L.*) cultivars. World Patent Information, 65, 102041.

[9]   Aidan, H., Julius, A., & Maurizio, G. (2003). *US Patent ID US6630507B1: Cannabinoids as antioxidants and neuroprotectants.*

[10] Marcu, J. P., & Schechter, J. B. (2024). The Role of FDA in Defining Cannabis and Hemp: A Brief Perspective on the future of Cannabis Regulation in the United States. In: Cannabis Innovations: Regulated Cannabis and Hemp Market Navigation, Business, Public Health, and Research Challenges. pp.15–40.

[11] Food and Drug Administration, Center for Drug Evaluation and Research. (2024, June 4). *Psychopharmacologic Drugs Advisory Committee Meeting (PDAC).* https://www.fda.gov/media/180703/download

[12] U.S. Patent and Trademark Office. (2019). *Examination Guide 1–19: Examination of marks for cannabis and cannabis-related goods and services after enactment of the 2018 Farm Bill.* (Accessed October 1, 2025) https://www.uspto.gov/sites/default/files/documents/Exam%20Guide%201-19.pdf

[13] European Union Intellectual Property Office. (n.d.). *Guidelines for examination, Part B.* (Accessed October 1, 2025) https://guidelines.euipo.europa.eu/binary/2213908/2005000000

[14] Health Canada. (n.d.). *Packaging and labelling guide for cannabis products.* (Accessed October 1, 2025) https://www.canada.ca/en/health-canada/services/cannabis-regulations-licensed-producers/packaging-labelling-guide-cannabis-products.html

[15] Government Gazette Republic of South Africa. (2024). *Cannabis for Private Purposes Act (Act No. 07 of 2024).* https://www.parliament.gov.za/storage/app/media/Acts/2024/Act_7_of_2024_Cannabis_for_Private_Purposes_Act.pdf

[16] Brazil. (1996). *Law No. 9,279 of May 14, 1996.* https://www.gov.br/inpi/en/services/patents/laws-and-regulations/laws-and-regulations/lpienglish.pdf

[17] U.S. Food and Drug Administration. (n.d.). *Small business assistance: Frequently asked questions for new drug product exclusivity.* (Accessed October 1, 2025) https://www.fda.gov/drugs/cder-small-business-industry-assistance-sbia/small-business-assistance-frequently-asked-questions-new-drug-product-exclusivity

[18] Marcu, J. (2020). The legalization of cannabinoid products and standardizing cannabis-drug development in the United States: A brief report. Dialogues in Clinical Neuroscience, 22(3), 289–293.

[19] Gosser, J., 2022. Comparative intellectual property protection for marijuana: United States vs. the European Union. Global Business Law Review, 11(78), pp.87–88.

[20] Willis, K. (2022). Avoiding the chaos of Maryjane: A conventional approach to intellectual property protection of marijuana. John Marshall Review of Intellectual Property Law, 17(278), 283.

[21] Cantore, C. (2025). The Economics Of Intellectual Property Protection for Cannabis: Challenges, Opportunities and the Evolution of Patents. In: Intellectual Property and Cannabis. pp. 272–293.

[22] Kamin, S., & Moffat, V. (2015). Trademark laundering, useless patents, and other IP challenges for the marijuana industry. SSRN. https://doi.org/10.2139/ssrn.XXXXXX (if known).

[23] Zimmerman, L., (2023). Budding opportunities: Cannabis trademarks in the United States. In: Intellectual Property and Cannabis. pp.16–28.

[24] Hutchison, C. (2024). The Canadian scenario. In: Intellectual Property and Cannabis. pp.201–204.

[25] European Medicines Agency. (n.d.). *Data exclusivity*. (Accessed October 2, 2025) https://www.ema.europa.eu/en/glossary-terms/data-exclusivity

[26] *Canopy Growth Corporation v. GW Pharma Limited*, No. 22-1603 (Fed. Cir. 2023). Justia. https://law.justia.com/cases/federal/appellate-courts/cafc/22-1603/22-1603-2023-04-24.html

[27] Sandell, L., (2024). Protecting cannabis cultivars in the United States: Utility patents, plant patents, and plant variety protection certificates. In: Intellectual Property and Cannabis. pp.162–164.

[28] Roberts, G. (2019). Beyond criminal prosecution: How federalist tensions in marijuana laws affect state-legal businesses. University of Florida Journal of Law & Public Policy, 29(413), 433–435.

[29] AbbVie. (2025, August 25). *AbbVie to acquire Gilgamesh Pharmaceuticals' Bretisilocin . . .* https://news.abbvie.com/2025-08-25-AbbVie-to-Acquire-Gilgamesh-Pharmaceuticals-Bretisilocin,-a-Novel,-Investigational-Therapy-for-Major-Depressive-Disorder,-Expanding-Psychiatry-Pipeline

[30] Kupferschmidt, K. (2024). FDA rejected MDMA-assisted PTSD therapy: Other psychedelics firms intend to avoid that fate. Science, 385(6710). https://www.science.org/content/article/fda-rejected-mdma-assisted-ptsd-therapy-other-psychedelics-firms-intend-avoid-fate

[31] Mimler, M. (2025). "A Bridge Too Far?": Trade Mark Protection for Cannabis-Related Products and Services in the European Union. In: Intellectual Property and Cannabis. pp. 40–60.

[32] Williams, Y. (2025). Geographical Indications and Cannabis. In: Intellectual Property and Cannabis. pp. 61–79.

[33] Goffe, M. (2025). Traditional Knowledge, Traditional Cultural Expressions and Cannabis: An Indigenous Rights-Based Access and Benefit-Sharing Framework. In: Intellectual Property and Cannabis. pp. 232–269.

[34] Corthésy, N. (2024). The Development of an IP Protection Strategy for Jamaican Cannabis. In: Intellectual Property and Cannabis. pp. 179–184.

Molly E. Burnett, Scott D. Greene, Andrew M. Peterson

Cannabis overdose: the myth of limitless safety

"The dose makes the poison." – Paracelsus

"I was gonna clean my room, until I got high (ooh-ooh-ooh)
I was gonna get up and find the broom, but then I got high (la da da da la da da da)
My room is still messed up and I know why (why man?), yeah hey
'Cause I got high'
Because I got high
Because I got high (la da da da la da da da)
I was gonna go to class, before I got high (come on, yo), check it out (ooh-ooh-ooh)
I could'a cheated and I could'a passed, but I got high (uh uh, la la da da da da)
I'm takin' it next semester and I know why (why man?), yeah hey" – Afroman, 2000

Introduction

"Can you overdose on cannabis?" This question has become more and more relevant as the cannabis legalization movement continues to expand across the United States and around the world as individuals gain access to higher potency products than ever before. The simple answer is – technically, yes one can overdose on cannabis. However, cannabis overdose presents significantly different compared to other drugs and rarely results in fatality.

The widespread belief that cannabis overdose is not possible stems from the decades of use demonstrating its remarkable safety profile compared to other controlled substances. Many individuals believe that cannabis is much safer than other substances, such as tobacco or alcohol, a perception bolstered by its origins as a "natural" plant and its historical use in various cultures [1].

Unlike opioids, alcohol, or stimulants, natural cannabis has never been definitively associated with a fatal overdose in humans when used in isolation. [2, 3]. This safety record has contributed to its reputation as a "harmless" substance, though this perspective oversimplifies a more nuanced reality. Scientific evidence indicates that regular cannabis use can lead to a variety of adverse health outcomes, including dependence or addiction, cognitive impairment, psychosis, anxiety, cardiovascular issues, and developmental problems when used during pregnancy [1, 4].

Terminology

What exactly is overdosing? How does it compare to toxicity?

"Overdosing" is a term used to describe pharmacokinetic overwhelm from a drug on the body's elimination pathways. Essentially, too much drug is introduced with not enough time and/or metabolic resources to biotransform and eliminate it from the body leading to toxicity, organ damage, organ failure, and sometimes death [5]. Toxicity can occur acutely (i.e., overdose or poisoning) but can also result when drug concentrations reach supratherapeutic range over a longer duration. When someone dies from a drug overdose, the medical examiner or coroner must indicate on the death certificate whether the overdose was intentional (a purposeful act, such as suicide) or unintentional (accidental). Unintentional drug-poisoning deaths include situations where a drug was taken by mistake, an excessive dose was taken unintentionally, the wrong drug was administered or consumed in error, or an accident occurred during the use of a drug in medical or surgical care. Therefore, while overdosing and toxicity are both consequential harm of excessive drug exposure, overdosing generally occurs as an isolated incident whereas toxicity is the result of drug overexposure whether acutely or chronically. See Table 1 for definitions and related terms.

Table 1: Definitions and related terms.

Intoxication	A disturbance in the brain resulting in altered behavior, mental status, and diminished motor control followed by ingestion of a drug or substance
Overdose	Toxic exposure resulting from intentional drug use (e.g., suicide attempts or purposeful misuse) or unintentional incidents (e.g., medication errors), with the latter classified as accidental overdoses
Poisoning	Injury or death caused by exposure to a harmful substance, characterized by clinical symptoms
Toxicity	The degree to which a substance can cause harm or damage to an organism

Moreover, overdose is more likely to occur when multiple drugs and/or substances are taken at the same time [6]. For instance, acetaminophen and alcohol rely on the same enzymes and metabolic pathways for elimination, thereby an increased risk for overdose and liver toxicity occurs when consumed in tandem compared to one substance alone.

Unlike most drugs, which follow predictable metabolic and elimination patterns, cannabis exhibits a variable half-life and complex pharmacokinetic behavior. The lipophilicity, variable metabolism, and prolonged elimination of cannabinoids create substantial interindividual variability in systemic exposure and effects. Consequently, factors such as route of administration, dose, frequency of use, age, genetic variation in metabolizing enzymes, and comorbid conditions determine the onset, severity, and duration of intoxication, toxicity, or overdose.

Intoxication refers to a disturbance in the brain resulting in altered behavior, mental status, and diminished motor control followed by ingestion of a drug or substance [7]. Unlike alcohol, which has specific blood levels to determine the degree of intoxication, cannabis intoxication is much more subjective and can occur outside of set blood concentrations. Factors such as individual tolerance, metabolic capacity, and method of consumption can influence the degree of cannabis intoxication.

However, cannabis requires an extremely high dose consumed over a short period of time to induce fatal overdose symptoms. For instance, as per the work of Turner & Agrawal in 2019, human trials to determine the lethal dose of cannabis do not exist, yet intravenous (IV) administration with a range of 40–130 mg/kg tetrahydrocannabinol (THC) has demonstrated death in animal trials [8]. For the average 70 kg person, this means consuming a minimum of 2.8 g of THC in a single session. Therefore, while not impossible, fatal overdosing on cannabis is highly improbable due to the large amount and small timeframe required to illicit the outcome. Especially when compared to drugs most commonly associated with overdose (opioids, methamphetamines, and cocaine), cannabis is generally considered safe.

Signs and symptoms of cannabis toxicity

The signs and symptoms of cannabis toxicity and overdose can look similar and may vary dramatically from one individual to the next, making clinical assessment challenging. Table 2 summarizes the principal signs and symptoms of cannabis intoxication. Individual responses to cannabis depend on multiple factors including route, duration, frequency, as well as age, sex, and genetic makeup [9]. This individual variability means that the same dose and preparation of cannabis may be beneficial for some individuals but produce toxic effects in others.

Table 2: Signs of cannabis intoxication.

Signs	Explanation
Ataxia	Impairment of coordination and balance resulting in unsteady movements or difficulty walking
Confusion and lethargy	Cognitive slowing, decreased attention, and reduced responsiveness to external stimuli
Disorientation and extreme fatigue	Impaired awareness of time, place, or situation, often accompanied by profound tiredness or somnolence

Table 2 (continued)

Signs	Explanation
Euphoria	Elevated mood characterized by elation, increased talkativeness, and impulsive behavior
Feelings of relaxation, openness, and confidence	Subjective sense of calmness and well-being, with increased sociability or emotional expressiveness
Gastrointestinal distress	Nausea and vomiting as the body attempts to eliminate excess intoxicant; in severe cases, may progress to loss of consciousness or, rarely, death
Loss of inhibitions	Diminished self-restraint leading to socially inappropriate or risky behaviors
Poor judgment	Impaired decision-making capacity, often resulting in engagement in hazardous activities such as operating a vehicle while intoxicated
Speech impairment	Slurred or slowed speech secondary to psychomotor impairment

Physiological signs and symptoms

Mild physiological signs and symptoms of cannabis intoxication typically include conjunctival injection (red eyes), mild tachycardia, postural hypotension, decreased muscle strength, and decreased hand steadiness [10]. The most common severe presenting signs include lethargy, hypotonia, hypoventilation, tachycardia, ataxia, and mydriasis, with vomiting and seizures also reported [11]. With doses above 7.5 mg/m^2 (300 mg or more in children), hypotension, respiratory depression, ataxia and delirium or even psychotic reactions may occur. More concerning physiological effects can include uncontrollable shaking, seizures, and unresponsiveness.

Cannabis hyperemesis syndrome

A condition known as cannabis hyperemesis syndrome (CHS) has emerged with the rise in recreational legalization and THC potency with estimates of 2.75 million people affected annually [12]. The exact cause of this condition is still under investigation but may be linked to the biphasic nature of cannabis dosing and/or the over stimulation and subsequent imbalance of the endocannabinoid system due to excessive THC [13, 14]. Typically, CHS presents with nausea, vomiting, and severe abdominal pain similar to cyclical vomiting syndrome but in association with chronic cannabis use. Symptoms of CHS temporarily resolve with hot showers/baths. However, these symptoms will continue as long as cannabis is being used, leading to complications like dehydration and electrolyte imbalances, which could result in seizures in severe cases [13]. The only true cure for this condition is to abstain from cannabis entirely.

There are three phases of CHS: *prodromal, hyperemetic,* and *recovery.* The *prodromal* phase is early on in the disease where mild symptoms emerge but may not result in any vomiting just yet. This phase can last for months to years and typically is not associated with any changes in eating patterns or compulsive bathing [14]. Moreover, the *hyperemetic* phase presents with severe and persistent nausea, abdominal pain, and vomiting that results in compulsive bathing in hot water to alleviate symptoms and causes weight loss and dehydration [14]. Finally, the *recovery* phase only occurs when cannabis use is fully stopped and symptoms resolved. Upon diagnosis, supportive therapy with IV fluids and antiemetics is the standard of care; eating and bathing behaviors should return to baseline during recovery.

Those typically presenting are early middle-aged individuals who have been consuming cannabis chronically beginning in adolescence. CHS is often associated with chronically consuming potent THC products, yet, not all chronic cannabis users will develop CHS. However, with the rise of cannabis use becoming increasingly prevalent and CHS remaining difficult to diagnose, chronic users should be aware of the potential risks.

Psychological signs and symptoms

Psychological symptoms range from mild euphoria and relaxation to severe psychotic episodes. Initial psychological effects typically include euphoria, time and space distortions, intensified sensory experiences, and/or motor impairment [10,11]. However, large doses may produce confusion, amnesia, hallucinations, anxiety, or agitation. Furthermore, cannabis intoxication can potentially lead to exacerbations of pre-existing psychotic diseases such as schizophrenia or even an acute psychotic episode.

Risks of psychotic episodes in persons with pre-existing conditions

Unfortunately, in some individuals with a genetic predisposition for psychiatric conditions such as schizophrenia, using cannabis can trigger psychotic episodes that cause the condition to emerge sooner than if the individual had abstained from cannabis. According to a study by Niemi-Pynttäri et al. [15], of 18,000 participants who experienced cannabis-induced psychosis, about 50% were later diagnosed with schizophrenia.

In addition to the Niemi-Pynttäri study, a growing body of research demonstrates that cannabis can substantially elevate psychosis risk among genetically or clinically vulnerable populations. Evidence from longitudinal and genomic cohort studies indicates that the interplay between genetic predisposition and heavy cannabis use is complex yet additive rather than purely interactive. A 2024 King's College London study found individuals with high genetic susceptibility to schizophrenia who used potent cannabis daily had the highest likelihood of developing psychosis, even after adjusting for genetic predisposition markers. Therefore, cannabis-related psychosis may manifest independently from genetic predisposition, despite both factors heightening the risk separately. Further, these data reinforce the notion that frequent users of high-THC cannabis face a substantially higher incidence of psychotic disorders, regardless of whether their genetic polygenic risk scores for schizophrenia were above the population average [16].

Not all cannabis is created equal

Cannabis products vary dramatically in potency, composition, and risk profile, significantly affecting toxicity potential. Cannabis preparations range from 0.5% to 5% THC on the lower potency end to high-potency concentrates containing 65–99% THC. Ultimately, the type of cannabis matters; the key to reducing harm and maximizing positive outcomes is an understanding of the cannabis product fully before consuming. Key factors to consider include chemical makeup, form/route of administration, and potency/dose. Keep in mind all of these factors coalesce.

Chemical makeup

As established, not all cannabis products are the same. This principle extends to the chemical makeup of the cannabis product and whether the product is entirely THC or if there are other cannabinoids present in the formulation. A product containing solely THC, like in the instance of a concentrate "dab" product or a high-dose ingestible product, is more likely to cause adverse effects and toxicity compared to one with a ratio of THC and other cannabinoids such as cannabidiol (CBD).

Furthermore, synthetic cannabis such as K2 and Spice are unregulated and therefore the true chemical makeup of any one of these products is unknown. Subsequently, fatal outcomes have been linked to synthetic cannabis consumption because they often contain mixtures of chemical byproducts and other drugs such as benzodiazepines and tramadol [17]. Synthetic cannabinoids can be more potent and carry a higher affinity for CB1 receptors compared to natural cannabinoids resulting in a highly variable yet more severe effect profile. Needless to say, synthetic cannabinoids should be avoided entirely.

Form and route of administration

The form and route of administration of cannabis plays a significant role in controlling drug exposure due to onset and duration of actions. For instance, when cannabis is inhaled, effects occur within seconds to minutes, providing immediate feedback regarding the dose. Whereas ingestible formulations take longer, about 45 min to 2 h, to produce a therapeutic effect and are subject to low bioavailability [18]. Orally ingested formulations are subject to first-pass metabolism, whereby THC's active metabolite (11-OH-THC) contributes to prolonged duration and systemic effects. Similarly, an inhaled product will have a shorter duration of action, up to 4 h, compared to one ingested, roughly 12–24 h [19]. Therefore, accidentally consuming too large a dose and having to wait out the effects for a while is far more likely to occur with an ingestible or edible dosage form compared to an inhaled formulation.

Even still, route of administration nuances among the same form must be considered before consuming to better anticipate outcomes. Using flower inhalation as an example, one should understand smoking carries more risk for long-term health due to combustion byproducts compared to heating and inhaling via flower vaporization devices. Moreover, cannabis oil and concentrate products are much more potent and therefore can deliver a significantly higher dose than expected if one is not careful. See Table 3 for common cannabis dosage forms.

Table 3: Common cannabis dosage forms, THC content, and recommended starting doses.

Method of administration	Dosage form	THC content (mg or %)	Recommended starting dose	Notes
Inhaled	Flower	15–36% THC (typically sold in 3.5 g or 7 g units)	1–2 inhalations; wait 10–15 min before repeating	Onset 2–10 min, peak 30–60 min, duration 2–4 h; effects dissipate faster than edibles
Inhaled	Vape cartridge or disposable	60–90% THC (300 mg – 4 g total)	1 inhalation; wait 10–15 min before repeating	Onset 2–10 min, peak 30–60 min, duration 2–4 h; high potency—use caution
Inhaled or oral	Concentrates (budder, RSO, wax, sand, etc.)	Up to 99% THC (0.5–3 g units)	Very small amount (1 inhalation or < 5 mg equivalent); use caution	Extremely high potency; inhaled onset similar to vape, oral onset slower; high risk of overconsumption
Oral	Gummies or other edibles	2.5–100 mg THC per piece (commonly 10–20 count packages)	2.5–5 mg THC initially; wait ≥2 h before additional dose	Onset 30–90 min, peak 2–3 h, duration 6–8 h; slower onset, risk of overconsumption if re-dosed too soon

Table 3 (continued)

Method of administration	Dosage form	THC content (mg or %)	Recommended starting dose	Notes
Oral	Beverage or tincture	2.5–50 mg THC per mL (typically 30 mL bottles)	2.5–5 mg THC initially; titrate slowly	Onset 15–60 min, peak 1–2 h, duration 4–6 h; sublingual absorption may be faster than standard edibles

When selecting dosage, route of administration, and dosage forms, individual factors as well as condition factors must be considered to maximize benefit and reduce unwanted effects. Individual factors such as body size, tolerance, and severity of condition help to determine starting dose, which can be titrated based on symptom control and tolerability. Certain conditions requiring abortive therapy such as migraines, nausea, and vomiting are better managed with inhalation formulations due to quick onset of action. In contrast, chronic pain, fibromyalgia, and seizures may be better treated with edible formulations due to the longer duration of action. From a medical approach, a combination of formulations will be the best way to mitigate symptoms as long as the lowest effective dose of each form is used. The same concept can be applied for recreational use: start low and go slow. The individual can always take a little bit more if the desired effects are not felt with the initial dose.

For edible dosing, somewhere between 2.5 and 10 mg is a reasonable starting point. There are edible products in the form of baked goods, gummies, troches, and capsules, and depending on the state, these can be dosed out in individual pieces. These are excellent for ease of measurement and those looking for lower potency formulations. Other products include tinctures, liquid resins, and Rick Simpson Oil (RSO), which require more forethought and care when administering because measurement is involved and formulations tend to be more potent. Tinctures can range from low (2.5 mg/ml) to high (50 mg/ml) potency making them excellent medical formulations for dose customization and titration. RSO, on the other hand, is dosed by one-half to one grain of rice sized amount and is at least 60% THC. Therefore, these products are reserved for individuals experienced with cannabis with a moderate tolerance as even on the lower end of potency calls for about 600 mg per 1 g.

On the other hand, cannabis flower or bud exists as plant material. Sizes for purchase range from a gram to an ounce with potency from 15% to mid-30% THC. Keep in mind that this varies from state to state based on rules and regulations. Using the rule of every 1 g containing 10 doses, a 3.5 g package of 15% THC flower holds 525 mg THC and 35 doses (15 mg THC/dose). A 1-g container of budder, a cannabis concentrate containing 80% THC, holds approximately 800 mg of THC, equivalent to about 10 individual 80 mg doses of THC. However, a miniscule amount of a budder can deliver a moderate dose of around 20 mg. Because such concentrates are difficult to measure accurately and can easily lead to overconsumption, they should be reserved for individuals with established tolerance or for those with severe conditions that warrant higher doses. By comparing these two inhalation formulations, one can determine an individual would need to consume a far greater amount of cannabis flower than concentrate to achieve the same THC dose. Thus, depending on the individual, certain inhalation formulations should be avoided to minimize over-exposure to THC.

Potency/dose

The dose of cannabis, particularly THC, is a major factor when considering whether or not an individual is likely to experience an adverse reaction from cannabis consumption. Typically, ingestible products report dose in milligrams and inhalation products report in percentage of cannabinoids. For instance, concentrate forms or "dabs", are usually 90% or higher THC, yielding 900 mg in a 1 g package and are designed to be used in very small quantities. Whereas flower, depending on the region, may contain 30% THC, which translates to a total dose of 1,050 mg THC in a 3.5 g package. Therefore, using extremely potent products can result in consuming a very high dose in a short period of time leading to a higher probability of adverse outcomes.

Ingestible products such as gummies or baked goods are labeled in milligrams which may make distinguishing the dose easier. Yet, considering edible products can range from 2.5 mg to 100+ mg per dose depending on the state and market, it is incredibly important for the end consumer to thoroughly read the product label to consume the intended dose. Edible packaging has improved significantly in recent years, with clearer labeling and standardized information. Still, consumers must read carefully to avoid unintended overconsumption. For example, the *Wonka Bar* shown in Figure 1 prominently displays "300 mg" on the front, which refers to the total amount of cannabinoids in the entire chocolate bar – not a single serving. However, the nutrition label on the back lists a serving size of only 25 mg (one piece), meaning each piece contains a much smaller portion of the total dose. This kind of labeling can be confusing, especially when attention-grabbing fonts and familiar branding resemble popular non-infused snacks. Without reading the fine print, a consumer might easily ingest several times their intended dose. Such products highlight the importance of careful label review and secure storage – keep cannabis edibles safely out of reach of children and pets.

Overall, understanding basic pharmacokinetic principles helps to determine which formulation and potency is the best fit for safe yet effective consumption. Always starting with a lower dose and gradually adding more as effects are felt is the best practice for avoiding side effects. Furthermore, from a medical or chronic use perspective, using the lowest dose possible to control symptoms is best in order to avoid building unnecessary tolerance and long-term side effects.

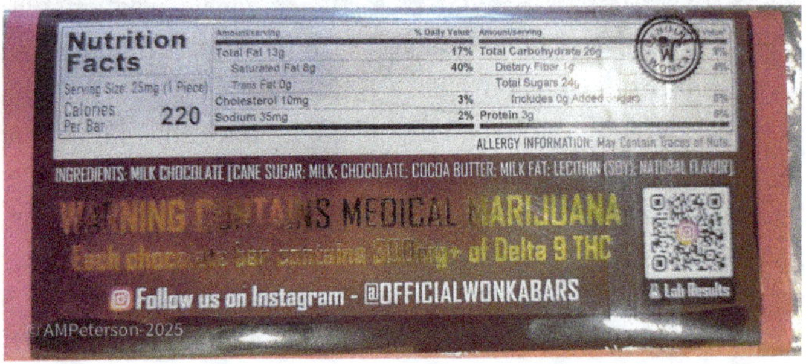

Figure 1: Front and back of a commercial cannabis-infused chocolate bar labeled "Wonka Bar." Photo courtesy of Andrew Peterson.

Quick Review

Cannabis toxicity presents with highly variable signs and symptoms that differ dramatically between individuals due to factors including route of administration, dosing, age, and genetics. Symptoms range from mild effects like red eyes and euphoria to severe manifestations including seizures, respiratory depression, acute psychosis, and exacerbations of pre-existing mental health conditions. Toxicity risk is fundamentally influenced by dramatic product variation – from low-potency flower (0.5–5% THC) to high-concentration extracts (90%+ THC) – and route of administration, with inhalation providing rapid onset compared to edibles' delayed but prolonged effects lasting 12–24 h. Critical safety considerations include understanding that form and concentration are primary determinants of toxicity, that THC-dominant products carry a higher risk than balanced THC:CBD formulations, and synthetic cannabinoids should be avoided entirely due to their unpredictable and potentially fatal effects. Likewise, the same dose may be therapeutic for one individual but toxic for another, emphasizing the importance of understanding product composition, proper dosing, and safe storage.

But is it fatal?

The assertion that one cannot fatally overdose on cannabis requires careful examination of the available literature, which presents a complex landscape of evidence quality and methodological approaches. While systematic reviews and meta-analyses of peer-reviewed studies provide the most robust foundation for understanding cannabis toxicity, much of our knowledge comes from less rigorous sources including poison control center reports, hospital admission data, and case series [20]. High-quality systematic reviews have utilized comprehensive database searches across platforms like PubMed, Embase, and the Cochrane Central Register of Controlled Trials to evaluate cannabis-related harms, yet even these gold-standard approaches often reveal studies with high or unclear risk of bias [21, 22]. The challenge is compounded by the voluntary nature of poison center reporting, which likely underestimates true exposure rates, and the limitations of hospital data collection systems which may lack specific substance information or dosing details [23].

The evidence base becomes particularly tenuous when examining non-medical cannabis, synthetic cannabinoids, and novel cannabis products, where knowledge relies heavily on self-report surveys, emergency department case series, and anecdotal reports rather than controlled studies [24, 25]. Retrospective analyses of cannabis-related hospital presentations acknowledge significant limitations including inability to rule out co-intoxication, missing data on consumption patterns, and the inherent biases of self-reported substance use. While databases like the Toxic Exposure Surveillance System and Drug Abuse Warning Network provide valuable population-level data, researchers consistently note that these sources vastly underestimate the total number of cannabis-related adverse events [21]. This methodological heterogeneity and variable evidence quality necessitates a nuanced interpretation of claims regarding cannabis safety, particularly the commonly held belief that fatal overdose is impossible with this substance.

Natural plant medicine

Natural cannabis and plant-based medicines may carry a reputation for safety and therapeutic benefit, but the reality is complicated by the kinds of products used, the degree of scientific evidence supporting their use, and how they compare to standard regulated pharmaceuticals.

Synthetic cannabinoids: a cautionary tale

Synthetic cannabinoids such as K2 or Spice present a sharp contrast to naturally derived products. Unlike regulated plant formulations, synthetic cannabinoids are often produced in uncontrolled environments with unpredictable ingredients and highly variable potency. This lack of oversight leads to significant health risks, including outcomes such as kidney failure and myocardial infarction. Furthermore, standard drug screens frequently fail to identify the specific synthetic agents responsible for these serious adverse effects, complicating both diagnosis and treatment [26]. Notably, most of these severe reactions lacked a clear antidote, and outcomes depended largely on aggressive supportive care. This unpredictability and the high rate of critical events set synthetic cannabinoids far apart from most plant-based cannabis preparations in terms of risk.

Comparing natural cannabis to established medicines

While the natural reputation of cannabis is attractive – especially compared with the side effects often seen with pharmaceuticals – the real-world risks and benefits are complex. Synthetic cannabinoids are public health threats with high morbidity and mortality, while medical cannabis demonstrates measurable (if usually mild) cardiovascular risks based on current evidence. Importantly, standardized pharmaceuticals benefit from robust safety data, precise dosing, regular quality oversight, and post-marketing surveillance – offering a level of predictability and control often lacking in unregulated cannabis products.

State-regulated cannabis, though safer than unregulated or illicit options, still faces notable imperfections. Variability in regulations across jurisdictions leads to inconsistent product quality, labeling inaccuracies, and sometimes significant contamination with pesticides or harmful microbes [27]. Weak enforcement and self-sampling practices can result in unreliable lab test results, with some states rarely testing retail products for actual cannabinoid content and contaminants. Consumer protection is further muddied by gaming of THC potency measurements and irregular standards for things like water content, which can degrade product quality before reaching patients.

Thus, seeking natural plant medicine may not guarantee safety or efficacy, and in some cases may expose users to greater risks than established, regulated medications, especially with untested, synthetic variants or potent extracts. State-regulated cannabis offers greater reliability and safety but does not meet the full standards of pharmaceutical rigor, leaving gaps that both consumers and clinicians should recognize.

In summary, while enthusiasm for natural medicines like cannabis is strong, both the actual product and the regulatory context matter profoundly in determining safety, and the benefits must be weighed carefully against the evolving landscape of risks relative to conventional medical products.

Evidence on risks

Plant-based medical cannabis, particularly standardized forms or pharmaceutical cannabinoids, has been studied for effectiveness and safety. In this section, we will review briefly review cardiovascular risks, car accidents associated with cannabis, and accidental poisoning risks. For more about the risks (and benefits) of cannabis in specific disease states, see chapter 10.

Cardiovascular risks

A systematic review spanning 47 randomized controlled trials with 2,800 patients revealed that medical cannabinoids used over a median of 16 days were associated with common cardiovascular effects – tachycardia, hypotension, and orthostatic hypotension [22]. However, serious cardiovascular events were not reported and most studies excluded patients with underlying heart problems. The data remains limited on long-term safety, especially in populations already vulnerable to complications. Cardiac effects like tachycardia and low blood pressure can be clinically relevant, but the current evidence suggests that these risks are generally of lower severity compared to synthetic alternatives, and far less than in other drugs.

In contrast, a more recent systematic literature review and meta-analysis showed habitual cannabis users – especially those using daily – are at a substantially higher risk for fatal events such as myocardial infarction (heart attack), sudden cardiac death, and stroke. Furthermore, these findings suggest a two-fold risk of death from cardiovascular disease when compared to non-users [28].

Car accidents due to cannabis

Due to the psychoactive effects of THC, cannabis can cause a decline in cognitive and motor function, as well as changes in time perception. Therefore, operating a vehicle or handling machinery is ill advised while under the influence of cannabis. Despite this warning, many cannabis users continue to operate a vehicle with cannabis in their system resulting in an overall uptick of collisions in states with recreational cannabis. Farmer et al. [29] concluded that "the combined effect of legalization and retail sales was a 5.8% increase in injury crash rates and a 4.1% increase in fatal crash rates." However, study findings regarding exactly how cannabis impacts driving have been mixed. With some studies demonstrating an increase in risk for crashes and others concluding cannabis makes no difference on driving capabilities when factoring in other risk factors such sex and age. Data from driving simulation studies by Brooks-Russell, et al. [30] and Hartman et al. [31] report slow reaction time, lane keeping, and road tracking decline along with decreased attention maintenance among cannabis-using participants. However, although increased following distance and slower speeds were observed, the extent to which these compensatory measures prevented crashes remained unclear. Likewise, a 26-study meta-analysis had inconclusive findings pointing to a 32% increased chance in crashes among cannabis users but without significant statistical data to bolster the collective claims [29]. And importantly to note, cannabis blood concentrations do not correlate with intoxication levels, which further confound incident rates related to cannabis use and vehicular crashes.

Overall, there has been a net increase in vehicular crashes following cannabis legalization, but to what extent varies from state to state with some reporting a 10% decrease and others a 4% increase [29]. Therefore, when using cannabis, one must use extreme caution and avoid operating a vehicle especially in instances where unexpected intoxication can occur such as trying a product for the first time, medical regimen adjustments (cannabis or otherwise) or dose titration periods.

Accidental "poisoning" due to cannabis (call to poison control)

Accidental poisoning due to cannabis most commonly occurs when ingesting edibles. Because edible products take longer to start demonstrating effects and have varying bioavailability, accidentally ingesting too high of a dose can be a common occurrence. Furthermore, edible cannabis packaging often features labels resembling familiar snack foods **(see Figure 1)**, particularly in recreational markets. This can lead to unintended overconsumption as discussed previously. Fortunately, most cases result in mild effects typically resolving with time and supportive care. However, accidental ingestion incidences involving children and the elderly are worrisome.

According to the Cannabis Poison Control System of California, between 2016 and 2023, the instance of calls related to cannabis exposure in children 5 years and under increased by 469%, followed by 147% amongst 6–19 year olds, and 68% of those 20 and older. Furthermore, 65% were calls in relation to children under five exposed to edible cannabis [32]. Another report showed that pediatric edible exposures in children younger than 6 years rose over 1,300%, with most cases occurring in home settings and a significant portion requiring hospitalization, including intensive care admissions. Children are especially vulnerable to the toxic effects due to their smaller body size and difficulty distinguishing edibles from regular food products, underscoring the need for safer packaging, robust parental education, and public health strategies to prevent unintentional exposure in household environments [33].

Conclusion

In conclusion, while the potential for a fatal overdose from natural cannabis remains rare, this chapter has demonstrated that cannabis toxicity encompasses a wide and complex spectrum of clinical presentations. From mild intoxication characterized by euphoria and motor impairment to severe and sometimes life-threatening conditions such as cannabinoid hyperemesis syndrome and acute psychotic episodes. The manifestations of cannabis-related adverse effects are multifactorial and influenced by product potency, mode of administration, individual susceptibility, and context of use. The substantial variability in cannabis formulations – from low-potency plant material to highly concentrated extracts and synthetic cannabinoids – significantly shapes the risk profile and clinical outcomes, underscoring the importance for user education, harm reduction, and careful clinical evaluation.

The evolving cannabis landscape poses novel challenges to clinicians, researchers, and policymakers alike. Despite decades of use and a relatively strong safety record compared to other controlled substances, gaps remain in our understanding of long-term toxicity, chronic effects, and the interplay between genetics and cannabis-related psychiatric risk. Public health initiatives must prioritize improving product labeling, standardization, and regulatory oversight to mitigate accidental overconsumption and pediatric exposures, which have increased alongside legalization and commercialization. Ultimately, a balanced approach respecting both the therapeutic potential and risks of cannabis use must be considered to minimize harm and maximize positive outcomes.

References

[1]   Bannigan, P., Evans, J.C., Allen, C. (2022). Shifting the Paradigm on Cannabis Safety. Leslie Dan Faculty of Pharmacy, University of Toronto. https://pmc.ncbi.nlm.nih.gov/articles/PMC8864419/

[2]   United States Drug Enforcement Administration. Marijuana. Washington, D.C.: U.S. Department of Justice, n.d. (Accessed October 13, 2025, at https://www.dea.gov/factsheets/marijuana.)

[3]   Rock, K.L., Englund, A., Morley, S., Rice, K., Copeland, C.S. (2022). Can cannabis kill? Characteristics of deaths following cannabis use in England (1998–2020). Journal of Psychopharmacology (Oxford, England), 36(12), 1362–1370. https://doi.org/10.1177/02698811221115760

[4]   Tacoma-Pierce County Health Department (TPCHD). (2023). Marijuana—Debunking the Myths. https://tpchd.org/wp-content/uploads/2023/12/Marijuana-Debunking-the-Myths.pdf

[5]   Platt, D. (1990). Pharmacokinetics of Drug Overdose. Clinics in Laboratory Medicine, 10(2), 261–269. https://doi.org/10.1016/s0272-2712(18)30567-5

[6]   Alcohol and drug foundation. (2021). Polydrug Use. Alcohol and Drug Foundation. https://adf.org.au/reducing-risk/polydrug-use/

[7]   Ernstmeyer, K., Christman, E. (2025). Nursing Mental Health & Community Concepts, 2e. https://wtcs.pressbooks.pub/nursingmhcc/chapter/14-2-substance-use-intoxication-and-overdose/

[8]   Turner, A.R., Agrawal, S.. (2019, June 3). Marijuana Toxicity. Nih.gov; StatPearls Publishing. https://www.ncbi.nlm.nih.gov/books/NBK430823/

[9] Kitdumrongthum, S., Trachootham, D. An individuality of response to cannabinoids: challenges in safety and efficacy of cannabis products. Molecules. 2023;28(6):2791. Published 2023 Mar 20. doi:10.3390/molecules28062791

[10] Kelly, B.F., Nappe, T.M. Cannabinoid Toxicity. [Updated 2023 Jul 10]. In: StatPearls [Internet]. Treasure Island (FL): StatPearls Publishing; 2025 Jan-. Available from: https://www.ncbi.nlm.nih.gov/books/NBK482175/

[11] Russo, L. Cannabinoid poisoning: Practice essentials, pathophysiology, epidemiology [Internet]. Medscape. Updated Jan 17, 2024 [cited 2025 Oct 13]. Available from: https://emedicine.medscape.com/article/833828-overview#a5

[12] I, A.M.. (2024). Cannabinoid Hyperemesis Syndrome. JAMA, 10.1001/jama.2024.9716. Advance online publication. https://doi.org/10.1001/jama.2024.9716

[13] Cleveland Clinic. (2021, July 25). Cannabis Hyperemesis Syndrome: What Is It, Symptoms & Treatment. Cleveland Clinic. https://my.clevelandclinic.org/health/diseases/21665-cannabis-hyperemesis-syndrome

[14] Sun, S., Zimmermann, A.E. (2013). Cannabinoid hyperemesis syndrome. Hospital Pharmacy, 48(8), 650–655. https://doi.org/10.1310/hpj4808-650

[15] Niemi-Pynttäri, J.A., Sund, R., Putkonen, H., Vorma, H., Wahlbeck, K., Pirkola, S.P. (2013). Substance-induced psychoses converting into schizophrenia. The Journal of Clinical Psychiatry, 74(01), e94–e99. https://doi.org/10.4088/jcp.12m07822

[16] Austin-Zimmerman, I., Spinazzola, E., Quattrone, D. et al. The impact of schizophrenia genetic load and heavy cannabis use on the risk of psychotic disorder in the EU-GEI case-control and UK Biobank studies. Psychological Medicine. 2024;54(15):4160–4172. doi:10.1017/S0033291724002058

[17] Cohen, K., Weinstein, A.M. (2018). Synthetic and non-synthetic cannabinoid drugs and their adverse effects-a review from public health prospective. Frontiers in Public Health, 6(162). https://doi.org/10.3389/fpubh.2018.00162

[18] Turner, A.R., Agrawal, S. (2020). Marijuana. PubMed; StatPearls Publishing. https://www.ncbi.nlm.nih.gov/books/NBK430801/

[19] Canadian Centre on Substance Use and Addiction. (2019). Inhaling vs Ingesting INHALING. https://www.ccsa.ca/sites/default/files/2019-10/CCSA-Cannabis-Inhaling-Ingesting-Risks-Infographic-2019-en.pdf

[20] Cooper, Z.D., Williams, A.R. (2019). Cannabis and Cannabinoid Intoxication and Toxicity. In: Montoya, I., Weiss, S. (eds.) Cannabis Use Disorders. Springer, Cham. https://doi.org/10.1007/978-3-319-90365-1_12

[21] Allaf, S., Lim, J.S., Buckley, N.A., Cairns, R. (2023). The impact of cannabis legalization and decriminalization on acute poisoning: A systematic review. Addiction (Abingdon England), 118(12), 2252–2274. https://doi.org/10.1111/add.16280

[22] Watanabe, A.H., Navaravong, L., Sirilak, T., Prasitwarachot, R., Nathisuwan, S., Page, R.L. 2nd, Chaiyakunapruk, N. (2021). A systematic review and meta-analysis of randomized controlled trials of cardiovascular toxicity of medical cannabinoids. Journal of the American Pharmacists Association: JAPhA, 61(5), e1–e13. https://doi.org/10.1016/j.japh.2021.03.013

[23] Shannon, M.W. Emergency Management of Poisoning. In Haddad and Winchester's Clinical Management of Poisoning and Drug Overdose. 2007;13–61. doi:10.1016/B978-0-7216-0693-4.50007-4

[24] Gresnigt, F., van den Brink, L.C., Hunault, C., Franssen, E., de Lange, D., Riezebos, R. (2023). Incidence of cardiovascular symptoms and adverse events following self-reported acute cannabis intoxication at the emergency department: A retrospective study. Emergency Medicine Journal: EMJ, 40(5), 357–358. https://doi.org/10.1136/emermed-2022-212784

[25] Kourouni, I., Mourad, B., Khouli, H., Shapiro, J.M., Mathew, J.P. (2020). Critical Illness Secondary to Synthetic Cannabinoid Ingestion. JAMA network open, 3(7), e208516. https://doi.org/10.1001/jamanetworkopen.2020.8516

[26] de Oliveira, M.C., Vides, M.C., Lassi, D.L.S., Torales, J., Ventriglio, A., Bombana, H.S., Leyton, V., Périco, C.A., Negrão, A.B., Malbergier, A., Castaldelli-Maia, J.M. (2023). Toxicity of synthetic cannabinoids in K2/Spice: A systematic review. Brain Sciences, 13(7), 990. https://doi.org/10.3390/brainsci13070990

[27] MacCallum, C.A., Lo, L.A., Pistawka, C.A., Boivin, M. (2023). A clinical framework for evaluating cannabis product quality and safety. Cannabis and Cannabinoid Research, 8(3), 567–574. https://doi.org/10.1089/can.2021.0137

[28] Storck, W., Elbaz, M., Vindis, C., Déguilhem, A., Lapeyre-Mestre, M., Jouanjus, E. (2025). Cardiovascular risk associated with the use of cannabis and cannabinoids: A systematic review and meta-analysis. Heart (British Cardiac Society), heartjnl-2024-325429. Advance online publication. https://doi.org/10.1136/heartjnl-2024-325429

[29] Farmer, C.M., Monfort, S.S., Woods, A.N. (2022). Changes in traffic crash rates after legalization of marijuana: results by crash severity. Journal of Studies on Alcohol and Drugs, 83(4), 494–501. https://doi.org/10.15288/jsad.2022.83.494

[30] Brooks-Russell, A., Brown, T., Rapp-Olsson, A.M., Friedman, K., Kosnett, M. (2019). Driving after cannabis use and compensatory driving behaviors among current cannabis users in Colorado. Traffic Injury Prevention, 20(sup2), S199–S201. https://doi.org/10.1080/15389588.2019.1665424

[31] Hartman, R.L., Brown, T.L., Milavetz, G., Spurgin, A., Pierce, R.S., Gorelick, D.A., Gaffney, G., Huestis, M.A. (2016). Cannabis effects on driving longitudinal control with and without alcohol. Journal of Applied Toxicology, 36(11), 1418–1429. https://doi.org/10.1002/jat.3295

[32] Cannabis Poison Control System Calls Dashboard. (2024). Ca.gov. https://www.cdph.ca.gov/Programs/CCDPHP/sapb/cannabis/Pages/Cannabis-Poison-Control-System-Calls-Dashboard.aspx

[33] Tweet, M.S., Nemanich, A., Wahl, M. (February 2023). Pediatric edible cannabis exposures and acute toxicity: 2017–2021. Pediatrics, 151 (2): e2022057761. doi: 10.1542/peds.2022-057761

Experts in the field interview with Lila McKinley

Ms McKinley has worked at the Department of Consumer Protection, which houses Connecticut's cannabis programs, for over a decade, from the start of the State's medical marijuana program, through the launch of the adult-use cannabis market. She was part of an adult-use cannabis transition team for the Department assisting in transitioning the State's medical marijuana program to a dual medical and adult-use cannabis program. Serving as the as the Division Director for the newly formed Cannabis Control Division within Consumer Protection since April 2025, she oversees Connecticut's adult-use cannabis and medical marijuana programs directing policy, enforcement, and licensing processes. Prior to that she served as a legal program director for the Department's Drug Control Division providing counsel in both medical and adult-use cannabis contexts and was an integral part of policy discussions and legislative initiatives. On a national level, she also serves as a board member and a co-chair on the Packaging, Labeling and Advertising Committee for the Cannabis Regulators Association (CANNRA). She earned her BA and JD from the University of Connecticut.

Jahan Marcu

Hello, I am Dr Jahan Marcu

Andrew Peterson

And I'm Dr Andrew Peterson, and we are the editors of the *Cannabis Innovation* series. And today we welcome another interviewee to our experts in the field discussions. This second volume of the series is focused on demystifying cannabis, where we examine the myths, mysteries and truths about various topics within the industry.

Jahan Marcu

We are here today to talk about research, science, and regulations. With us today is Lila McKinley, director of the cannabis Enforcement Division for the Connecticut Department of Consumer Protection. It is always good to talk with someone who works in the regulatory side of the industry, particularly with someone who has such a long standing experience overseeing a cannabis program with 10 plus years of tenure in the field. So, please share with us what's your vision for Connecticut's industry and how has that changed over the years

Lila McKinley

Sure. So cannabis is a very still…I would say the industry is very nascent. You know, when Connecticut started in this particular industry legalizing medical cannabis, it was a very different climate. I think the focus was very much on medical, we engaged in a very medical/pharmaceutical model. And as we saw over time, states slowly beginning to legalize the adult use – or some, some people refer to it – the recreational market. We eventually, in Connecticut, legalized the recreational market as well, and we saw a shift in sort of the different policy choices that we had to make within our regulatory structure, because not only were we housing a medical market, but we were also housing a market that was serving just the general public for adult use and recreational market. So really, what has shifted over time is figuring out how to balance both the medical market with the adult use market, ensuring both public health and safety but also keeping both markets alive, flourishing, and again being safe for consumers and medical patients.

Andrew Peterson

Wow. There's a lot to unpack in there. One of the questions that I have, though kind of relates a little bit, you know, how does the focus on regulations and policies change, and is that you've been reflecting back on what the consumers want, what the legislator wants. How's that working?

Lila McKinley

So it is...that's a very broad question. It you know, it changes over time with the different players, I think that come into the market space as well. So, you know, our regulations have always, I do want to underline that our regulations, since the beginning of our foray into the cannabis market, whether medical or adult use, they've always been the underlying basis for us, because we are the Department of Consumer Protection which is all about public health and safety, and so that is always at the forefront whenever we're crafting any sort of regulation. I think the challenge and the shift comes in is when you are in a medical market, what public health and safety means, is a little bit different than in when you're in the adult use market, because you are approaching it from the concept of so this is someone's medicine, and you want to make sure that their medicine is safe and effective for what they need. Whereas on the adult use side, you're focusing on a product that someone is using for recreational purposes. I think what we've seen over time, however, when you talk about the role of people coming into the legislature and who's crafting the rules and the regulation is that we've seen a real shift from medical when we moved into adult use of just generally, more people having interest in the regulatory scope. So, the legislature has become more involved in crafting legislation around it. Our governor's office has also become more involved. Other agencies have become more involved. Because we are touching them more right? We're talking about taxation, so we're touching the Department of Revenue. We're talking about enforcement and the criminal side for cannabis violations for illicit stores. So we're talking about our Department of Public Safety, so and then Department of Public Health to do some of our research things. So what we're really seeing is many more agencies coming into the fold.

We're also sort of seeing the legislature taking a much more engaged approach and having more conversations with public health advocates as well as industry advocates. And then we're just seeing a lot more players coming into the space, whether it be industry or public health advocates or people that are providing adjacent services to the cannabis industry. And I think part and parcel of it is because, obviously we're expanding the scope of the marketplace in engaging in adult use, but also I just think the cannabis industry itself is expanding, so we are seeing that growth there, and people are paying a bit more attention as it sort of expands out nationally. So, when I started 10 years ago in the medical space, there certainly weren't as many voices and engaged folks as there are I think now.

Jahan Marcu

Wow, I think the public health angle is really important as we all are sort of watching in real time as new products are coming out, new formulations, new ingredients, and just ways that consumers are engaging with these products. So I have to wonder, how does your office and your colleagues balance promoting what's best for public health that can be anything from, you know, underage prevention of use, more rigorous testing standards, more public health messaging, with creating a viable market where people want to use sort of any product that can be made from the cannabis plant?

Lila McKinley

It's a very fine balance, because certainly, and I think to some degree, having a safe market space benefits the industry overall, right? Because it creates consumer confidence in the market space. And so, I do try to take that approach with our industry as well, to have a bit more of a collaborative approach to recognize that the more that we can give legitimacy and make our consumers feel safe and confident in the market space and that they're being protected, the more that we can see industry growth, because people will be more apt to engage in the market. So I think that's always the underlying commonality that I try to find when we're trying to work together to get an appropriate policy. I think the public health piece is also a bit challenging because of the way that the market space evolved. So when we talk about the initial legalization, I think, you know, really the focus there was just on the legalization and as the market has developed and as new states have come on board, I think we've seen the shift of that focus to be a bit more public health oriented.

For us, our approach has always been to it's very hard when you're a regulator to come in and prohibit something that's not prohibited previously, that's a big challenge. So really, I think for us, the approach has been to start out conservative and see where that's sort of taking us, and if we do not see public health impacts, and we think that we can maybe loosen that regulation to help industry kind of get a better foothold in the market space. Then we'll continually look at that and then maybe roll back some of those very much more conservative laws that were focused on public health and safety. I think the other thing too, is sort of making sure that you have the floor. So there are definitely regulations and issues that we know that have come up in this space, whether it be the need for testing product or making sure that packaging and labeling doesn't appeal to children, accidental ingestion, those types of things. There are things that most all states do to really hone in on those very significant public health and safety issues. And so, we always make sure that we are addressing the floor and starting from that, and then kind of going up with a bit more of a conservative approach. But it's, it's, you know, it's a challenge, because I think there's a desire for people to want to be successful in this industry, and that's often what they will tell the legislature that they you know, this regulation is preventing me from doing business. It's hurting my business.

But then on the flip side, it's also making sure that the consumer is safe and making sure that the market space is safe. And so, you know, if we know that a certain testing, for example, for heavy metals in the product, we know it's pretty well supported that heavy metals are dangerous, particularly for the methods of ingestion and cannabis, such as inhalation. And so, it's really important that we educate the legislature around those, you know, very important aspects, and then if there are things that are a bit more stringent that we realize, well, okay, so maybe, you know, we can come at this a little bit different of a way to still protect public health and safety, but let the industry sort of have a little bit of leeway to do what they need to do. That's when we are able to kind of toggle back and work within that space. But it's definitely always a challenge. I think it's a challenge most industries face, and it's something I think that at least in this industry, there's so many new players, it's sort of a new space that everyone's sort of shocked at first, but, you know, it's not anything necessarily new when you're trying to regulate.

Jahan Marcu
I do have a follow up to that, and it makes me think about how difficult it must be to make those decisions, especially if you're trying to be conservative and not, you know, prohibit something that is allowed and things like that. What sort of research or data are your colleagues working on, like in the Department of Health? Or you could take it another way. What do you wish they could study? Are there other research questions where you like I want data to answer this question, or if there are projects that are ongoing that you could share a little bit about that would be great.

Lila McKinley
Yeah. So, I think the particular challenge in the cannabis space is the federal illegality, right? And so, there is a lack of research and funding that comes from the federal side, and I will say our agency does regulate other well established industries, like pharmaceuticals, for example. So, we're very we're very sensitive to that, because we can see a side-by-side comparison of sort of the funding and the things that the pharmaceutical industry is able to do because we regulate it, as opposed to cannabis. I think how we make our decisions is we try to make the most informed decision with the data that's available, and sometimes that data can be limited. But we also will look to other industries as well. I know it's very common to look to the liquor industry. It's very common, particularly for us. We have our agency actually holds a ton of what I'd like to call the vices. So, we regulate everything from gaming to other

licensing functions. So, we will also look to different areas, you know, tobacco, nicotine, anything that can help inform us of making decisions. We rely greatly on other states, and also there are resources in those states. So, for example, when we were putting in place our testing regulations, we did a lot of consultation with microbiologists that were on staff in other cannabis agencies in other states. And then just sort of another example of kind of like moving outside of cannabis, we also spent time with microbiologists that were in the pharmaceutical space, not necessarily the cannabis space, to help inform those regulations.

In terms of what we're working on and what we wish we had for data. It's just there's so much our Department of Public Health participates in a lot of survey data, so a lot of surveying of consumer usage and the like. I think, for the regulatory scope, what we are really missing, and what I think we would love to have, is things about, like the effectiveness of the regulation. So, I'll just take for example, like child protection and packaging and labeling. There are a ton of requirements from state-to-state about what can appear on the package and label. There are requirements about warnings, health warnings, some of those have been borrowed from other states that are, and other countries that have been, involved in that research, like Canada, for example. But really, there's not a ton of data on how effective those regulations we have are on the consumer experience. So, you know, there isn't a ton of research done on whether or not I don't know a cartoon appeals to children, or if a particular warning label about cannabis is actually, you know, supported by data. And even something as simple as the universal symbol, which appears on our cannabis packaging, whether or not that's really conveying the message to folks that this is something that contains cannabis they should keep it out of the reach of children and prevents accidental ingestion. So, we don't have a ton of data around those types of things, which I think we would love to see more of, because it sort of gets back to the question that you had about balancing public health and safety and regulating. And the more data that we have to inform those decisions, the more we can effectively put that balance into place and make sure that a), we're doing what we need to do to protect public health and safety, but b), we're not being overly conservative when we're doing it.

Andrew Peterson

Wow, I like that. I'm a researcher, and I'm interested in data. Part of me wants to know, where would you get that data? What are the challenges of getting it? And money is probably one of the bigger issues. Trying to figure out where you're going to get funding and be able to collect that data. And then how do you translate some of even if you did it in Connecticut? How does New York, New Jersey, or Texas, or not necessarily Texas, but any of these other states may have something going on. How do they use it too? So it's, it's a lot and a lot of entangling things. And I bring up the other states because I want to transition us a little bit into, kind of, given some of the experiences that you've had within Connecticut, what lessons or recommendations would you give to other states who are considering, you know, launching a medical cannabis program, or an adult use or if you know in their vision is to go from medical to adult use, kind of what are some of the lessons that you've learned that you think they would benefit from learning from hearing from you.

Lila McKinley

So, Connecticut had the benefit of going a little bit later when it came to legalizing adult use we certainly weren't one of the first, but we're definitely not one of the last. And I think as each state comes on board, we're able to learn lessons and improve every time someone enters into the adult use space, for the most part. I think we've learned a lot. I think there are a few key sort of takeaways, things that maybe distinguish us a little bit from some of the older states that came on board. I know that the older States didn't really have social equity programs. I think that's been a big component for newer states programs, and it was for ours. So, we, you know, effectively, do have a licensing scheme for our social equity folks, but we also learned from Massachusetts who just had a licensing scheme and not necessarily funding sources and things that actually went back to the community to help the community that was most impacted by the war on drugs. We also have reinvestment provisions, so provisions where some of our licensing fees and funding from the tax, the taxes that we collect, are actually going back into those communities. And so, I think that is actually one of the more important lessons that we've learned as it relates to social equity. Because I think the initial focus was just on giving them a license in the cannabis space. But that only helps a few, it doesn't necessarily help the entire community. And so, I think that was really a big cornerstone for us, and obviously rolling back convictions.

The public health and safety side, I think we learned that a conservative approach is significantly better for consumer experience. So, we do have some pretty stringent packaging and labeling requirements, particularly around our edible market, such as just plain black and white packaging to make sure that we can prevent accidental ingestion. We've also learned a great deal about lab testing, so we continue to kind of build off other states and enhance both the things that we test for, but also just get more specific and how we're testing for things in the methods that are employed.

And then I think, generally, you know, the other big thing that we've learned is making sure that, like all of the that we'll never get as much, because, you know, funding and money is obviously a big thing for us, we'll never get as much as we would love, as any government agency would, but making sure that we have the appropriate funding and the appropriate staff to be able to kind of carry out some of those regulatory functions, and making sure to ask for the appropriate funding, and also just Having a barometer of, like, what that funding is.

I think also just overall engagement and education. We've learned that education and the sources of education are very important. So, whereas there are some industries where people implicitly will trust what the government is saying. In this industry, sometimes you need to have different voices conveying information besides just the government. So, working through that communication plan, as well as another thing we learned that, you know, we may be trying to send messages, but we're not effective in getting those messages out, because we aren't we aren't the right people to say those messages to the target audience that we're trying to get out.

So those are some things we've learned. And obviously preserving the medical program, I think quite a few states in the West didn't really have an eye for that, and Connecticut is unique in that it's always had a pharmaceutical model, which means it always has a pharmacist on staff within our dispensaries, and kind of follows that model of your get your medication from a pharmacist, and so trying to keep that in place and just be cognizant of, you know, having that medical market and having that support for our medical patients, because for them, this product is their medicine, and it's used to treat conditions. And so it's really important that we make sure that that market stays viable, and prioritizing that over maybe recreational market spaces.

So we've learned, we're still learning, but we've learned, you know, from other states, things that they've tried in order to preserve that market space. We've tried a couple of new things, I think, kind of going forward – you know, advice for other states is just to sort of keep building on what newer states in the market have done definitely go back and talk to those states, and they will always tell you what they wish they could see. And it's not a perfect thing, because despite what you learn there, you're also making regulations in an environment where there are legislators and advocates, and so sometimes even though there is a better practice that, at least that you think could be put in place, you still have to sort of contend with those other viewpoints. So sometimes states still do kind of revert back to some of their some of the older things that we've seen, but we have seen that states have really been working to prioritize their medical programs.

And the other thing is some new states. Everybody starts at a different spot, I think, which is really the interesting part. So, where Texas is in terms of their cannabis market is very different from, say, where Connecticut was, or even just as simple as, like as where Vermont was. And so everyone sort of starts in a different space. So there you can learn lessons, but you also sort of have to go back to your state and see where you're starting from, and if those lessons are going to be applicable to you

Jahan Marcu
Absolutely. And I imagine that dealing with everyone starting in a different space, when you meet with your colleagues like at CANNRA and other organizations, and you see where they're at, you're like, Oh, I wish I had known X or Y if I was in that situation. And so, you know, you talked a lot about information and how important it is to communicate and share messaging. What are resources that you find yourself just sharing with people. Are there books, podcasts, social media platforms where

you're like when you feel stuck in the cannabis regulatory world? I look to hear for inspiration.

Lila McKinley

So I think you know, there are definitely, we have our own resources that the agency puts out. So, we try to make them very simple, we've dabbled in videos with like, little cartoons. We also have a ton of, you know, frequently asked questions and the like. But where do where do I look? Well, first of all, I, for sure, look to CANNRA. So that's the cannabis regulatory association. So it's like all of the states that are involved in cannabis space are part of the organization, and there are a lot of documents that are shared there. You know, the state, the state might have done a study, and they'll post it for us to see. Or, you know, health, there was recently a health study on CBD, and that was something that CANNRA made us aware of, so that's probably the primary source. But I also, you know, oddly enough, there are news outlets like MJ Biz that provide updates on cannabis issues. And I like the articles because it will kind of let me see Missouri is doing this, or New Jersey just did this. And so then I can, like, flag it and maybe talk to the folks in New Jersey and say, "Hey, I see you've done this new thing. I'm really interested in it." So that's, that's another place I think that's super helpful. And then sometimes just standards organization too. So USP puts out a ton of standards and has been working on some things related to cannabis, as well as ASTM and so looking at those resources and figuring out where, you know they're looking to make standards, what their standards look like. That's also very helpful and very relevant, but really other regulators. I mean, if you want to know how I we decompress and de-stress, it's talking to other regulators in other states, because they are similarly situated and often face the same challenges, so they're uniquely positioned to understand the struggles.

Andrew Peterson

So it's a camaraderie of some sort, yeah, yeah, that's nice. So very, very happy to hear that. And, and, you know, like us, you know, I go to other pharmacists, other researchers and stuff to talk about some of the struggles that we're having, too. So, I certainly can relate to that cool, nice. And MJ biz. That's a good one for us. You know, I take a look at that pretty regularly. Keeps me up to date on a lot of things. So, and then I do listen to some podcasts like Jahan does, I'm sure. And Jahan actually has some podcasts too.

Jahan Marcu

I definitely have been finding myself drawn to Marijuana Moment quite often, and then looking for articles on social media platforms and reading the comments to see how people are responding to this information. And that is entertaining, but sometimes there are some really surprising, insightful comments from the public. And so, I definitely think that it's always great to get the public opinion on something and just throw it out there, but it can be difficult.

Lila McKinley

Yeah, we will look at and review public commentary generally and if we have social media posts we will review those too. And, I mean, that's not to say, I mean, obviously, because we're the government there, there can be some negative connotations in those comments. But, you know, constructive criticism is always helpful. And it's not to say that, you know, I think sometimes people get very put off by a negative comment, but there can be things that are in there that you're like, oh, wait a second, you're right. The regulatory rule does have this other impact that was an unintended consequence. Is there a way to recraft the rule so that we can, you know, not have that unintended consequence, but still sort of put out, have that purpose of what the rule was intended to do. So as much as criticism hurts and stings, I also find it to be very important for us as regulators, particularly because there are a ton of voices, and it's very easy to just sort of filter them all out, because there's a lot of voices out there, but nevertheless, I think some of those voices can be very instructive on policy and just some of the things that we're doing great.

Andrew Peterson

Lila, this has been absolutely informative. Really appreciate you taking time out of your busy schedule to chat with us today about some of the challenges and some of the opportunities and strengths that you've had as the lead regulator for Connecticut. So really, so happy to have you here with us today. Thank you very much for joining us.

Lila McKinley

Thank you for having me. Appreciate it. No problem.

—

MYSTERIES

Mariana Larrea Arias
Prohibition's afterparty: fear, stigma, and modern cannabis policy

Introduction

Cannabis has traveled an extraordinary journey throughout human history, from being a valued medicinal plant in ancient civilizations to becoming the object of systematic demonization during much of the twentieth century. The "Reefer Madness" campaign in the United States during the 1930s marks a milestone in constructing the modern stigma around cannabis, a narrative that was exported internationally with devastating consequences for Mexico and other Latin American countries, which became the main stage for a governmental response to security concerns that has cost hundreds of thousands of lives [1].

Having witnessed firsthand the evolution of cannabis policy in Mexico, I can attest that this paradigm shift has been particularly complex. The plant that ancient Mexicans used for medicinal, industrial, and ceremonial purposes was criminalized under external pressures during the twentieth century [2]. Faced with escalating security threats and lacking alternative policy tools, the Mexican government adopted increasingly militarized approaches to drug trafficking. While this strategy succeeded in disrupting some criminal organizations, it also generated unintended consequences, including the fragmentation of cartels into smaller, more violent groups. However, it's important to recognize that these policies represented the primary policy instruments available to governments at the time for addressing organized crime. The twenty-first century has brought about a global reconsideration of prohibition, driven by lessons learned from implementation challenges and by growing scientific evidence regarding the therapeutic benefits of the plant.

What many fail to recognize is that the prohibition of cannabis, particularly industrial hemp, was never truly about public health concerns. As economic and historical records reveal, powerful wood and cotton industry interests in the United States actively lobbied for hemp prohibition in the early twentieth century, fearing competition from this versatile crop. This economic motivation, coupled with Nixon-era racial targeting, created the foundation for decades of stigma that continues to influence tax policies, banking regulations, and economic opportunities in the cannabis sector today.

At a time when the world is transitioning from prohibitionist paradigms toward diverse regulatory models, understanding the technical, social, and economic nuances of this process is fundamental for policymakers, industry stakeholders, and civil society interested in contributing to an informed debate about the future of cannabis.

© 2026 Walter de Gruyter GmbH, Berlin | https://doi.org/10.1515/9783111476162-006

Cannabis definitions and regulatory models: the technical foundation of policy choice

In the realm of cannabis regulation, definitions are not mere semantic matters but fundamental tools that determine the scope and limitations of entire regulatory structures. As American jurist Oliver Wendell Holmes Jr. aptly expressed, "a word is not a crystal, transparent and unchanging; it is the skin of a living thought and may vary greatly in color and content according to the circumstances and time in which it is used" [3]. The way legislation defines "cannabis," "hemp," "cannabinoids," and other related terms establishes the foundation for the entire legal and commercial ecosystem, determining which plants can be legally cultivated, which products can be marketed, and which activities are permitted throughout the value chain.

How definitions shape regulatory frameworks and market structure

The definitional choices made by regulators create entirely different economic and social realities. A single plant – *Cannabis sativa* L. – has been artificially divided into distinct legal categories based primarily on the content of one compound: tetrahydrocannabinol (THC). This division, although practical from a regulatory perspective, ignores the botanical and phytochemical complexity of the plant, which contains more than 100 different cannabinoids and hundreds of other bioactive compounds.

What makes this definitional challenge particularly complex is the plant's remarkable versatility. Cannabis can serve to build houses (construction materials with hemp fiber), feed families (protein-rich seeds), produce personal care products (oils and cosmetics), replace polluting plastics (hemp-based bioplastics), and provide therapeutic relief (medicinal cannabinoids). However, restrictive and fragmented definitions artificially limit this potential, creating regulatory silos that ignore the interconnected nature of the plant's applications.

The legal definition of "cannabis" determines which plants are regulated and how they are classified. If a definition is based solely on the percent of THC levels, it creates a binary system and confusing rules that establish different standards for cultivation, processing, distribution, and consumption. These seemingly technical choices have profound practical implications: they define tax structures, banking access, import/export procedures, and criminal penalties. In emerging markets, I've observed how minor definitional distinctions create entirely different investment landscapes and business opportunities.

Hemp versus cannabis: the THC threshold dilemma

The arbitrary nature of THC thresholds becomes apparent when examining global variations. The United States established a 0.3% THC limit in the 2018 Farm Bill, and after 7 years, in November 2025, the government announced step back and prohibiting the intoxicating hemp derivatives at the federal level. Canada uses different thresholds for different applications; the European Union recently increased its limit from 0.2% to 0.3%, and Mexico established a 1% threshold. These variations are not based on scientific consensus about psychoactive effects or public health risks, but rather on stigma, political negotiations, and regulatory convenience. This creates difficulties on all the levels, from the genetic of the seeds and the cultivation to the extraction and final use in products, making companies to expend thousands of dollars washing flowers and creating strategies to comply with the thresholds established by their regulations.

In Mexico, the General Health Law, modified in 2017, establishes this 1% THC threshold, making an exemption for industrial uses [4]. This opened opportunities for companies to import, export, manufacture, and commercialize products containing THC as an ingredient, like cosmetics, food, beverages, and dietary supplements. However, Mexico does not establish any regulations in favor of or against semi-synthetic cannabinoids as the United States does, because this is not a legal definition under the Mexican legal framework. Companies looking to import intoxicating hemp derivatives should therefore implement a legal strategy to confirm if those products comply with the Mexican legislation. On the other hand, countries as the United States, Czech Republic, Japan, and Colombia – each with considerably lower THC limits – are changing regulations and reducing thresholds, or banning other substances as hexahydrocannabinol (HHC), cannabidiol acid (CBDA), and cannabinol.

The thresholds are not what regulators should be worried about when thinking about the industry. Those percentages could be manipulated in a laboratory. Instead, the focus should be on education and building safe platforms for consumers. These platforms can help consumers understand what these cannabinoids can do, how they affect their bodies, and how they can use them more effectively to obtain maximum benefits from them.

The psychotropic versus psychoactive distinction in Mexican law

One of the most significant challenges in Mexican cannabis regulation involves the confusion between "psychotropic" and "psychoactive" substances. This distinction has profound practical implications for product classification, import procedures, and regulatory compliance strategies.

Under Mexican law, "psychotropic" substances are specifically listed in the General Health Law and subject to strict medical and pharmaceutical controls. "Psychoactive" substances affect mental processes but may not require the same level of restriction. Cannabidiol (CBD) in Mexico is regulated as a psychoactive substance under the Medical Cannabis Sanitary Control Regulations, but it is not classified as psychotropic, creating important legal pathways for industrial hemp derivatives [5].

This distinction becomes particularly relevant for semi-synthetic cannabinoids derived from industrial hemp. While compounds like delta-8-THC or HHC may produce psychotropic effects similar to THC, their legal classification follows their source material rather than their physiological effects. From a regulatory strategy perspective, this means such products can potentially be imported and commercialized under the psychoactive rather than psychotropic pathway, provided they comply with the 1% total THC limit.

International approaches and their regulatory consequences

Divergent definitional approaches worldwide have created remarkably different regulatory landscapes, each with distinct opportunities and challenges for businesses and consumers.

The US's 0.3% THC definition paradoxically opened the door to a parallel market for psychoactive cannabinoids derived from hemp, such as delta-8-THC and HHC [6]. These compounds technically comply with the legal definition but generate psychoactive effects similar to traditional cannabis, creating a regulatory loophole that has spawned an entire industry focused on intoxicating hemp derivatives.

Canada opted for a more integrated approach, regulating both cannabis and hemp under the same legal framework with differentiated requirements based on THC content and intended use [7]. This has enabled more coherent market development and avoided some of the regulatory contradictions observed in the United States, although it has also created more complex compliance requirements for businesses operating across multiple product categories.

The European Union maintains a hemp definition anchored in THC content. Still, it takes a more restrictive approach to derived cannabinoids, explicitly limiting the marketing of products with psychoactive effects regardless of their botanical origin [8]. This approach seeks to close the regulatory "back door" that has allowed gray markets to flourish in other jurisdictions, but it has also limited innovation and market development in the cannabinoid sector.

These definitional differences create distinct business environments: while the United States has seen explosive growth in hemp-derived intoxicating products, EU companies have focused primarily on nonpsychoactive industrial, cosmetic, and nutritional applications. For Mexico, whose economy depends significantly on international trade, these variations present both challenges and opportunities for positioning in the global market.

Medical versus recreational frameworks: models of regulation

The global evolution of cannabis regulation has generated a living laboratory of public policies, with various models reflecting different social, economic, and cultural priorities. These approaches can be categorized into four primary models:

The medical-only model, adopted by countries such as Germany (until recently), Australia, and Thailand, allows access to cannabis only with medical prescriptions under strict professional supervision. This approach prioritizes scientific evidence and quality control but may limit access and maintain parallel illegal markets.

The decriminalization model, implemented in Portugal and partially in Mexico through jurisprudence, eliminates or reduces criminal penalties for possession while keeping production and distribution illegal. This reduces harms associated with criminalization but does not address security, quality, and taxation issues related to black markets.

The dual regulation model, adopted in Canada, Uruguay, and several US states, establishes separate but complementary frameworks for medical and adult use. This approach recognizes different market needs and expectations, though it can also generate administrative complexities and regulatory discrepancies. A clear example is the conflict between the Marijuana MSO's (Multi-State Operators) in the United States and the hemp derivatives industry, which has been regulated since 2018, with thousands of dollars invested in lobbying and networking efforts to win this regulatory battle.

The integrated model, emerging in jurisdictions like Malta, establishes a single regulatory framework with variations based on end use, without creating parallel systems.

For Mexico, the choice between these models must consider both domestic social realities and international lessons learned. The ongoing debate reflects broader tensions between different visions of drug policy, public health, and individual rights [9]. Proponents of strict separation argue that medical cannabis should meet pharmaceutical standards and access to patients, while adult use should focus on consumer education and harm reduction. Those favoring integration point out that the use distinction is often artificial, as many consumers use cannabis for both medical and recreational purposes.

Emerging challenges: biotechnology and synthetic cannabinoids

The regulatory dilemma deepens with advances in biotechnology that allow the production of specific cannabinoids through fermentation or chemical synthesis, without cultivating the plant. These developments challenge traditional plant-based definitions and create new regulatory gaps.

How should synthetically produced CBD, THC, or novel cannabinoids be classified? Are they "cannabinoids" in the traditional legal sense? Current definitions rarely address these emerging issues, creating regulatory uncertainties that the industry inevitably explores. Some jurisdictions are beginning to regulate cannabinoids regardless of source, while others maintain distinctions based on botanical origin or THC percentages.

The emergence of intoxicating hemp derivatives, or novel psychoactive cannabinoid compounds that don't occur naturally but can be synthesized from hemp-derived precursors, presents particular challenges. These substances can circumvent THC-based regulations while producing similar or even stronger psychotropic and psychoactive effects, highlighting the limitations of current definitional frameworks.

This biotechnological evolution and the demand of consumers suggest that future cannabis regulation may need to shift from plant-based definitions toward compound-specific approaches, and the final purpose of each product, regulating individual cannabinoids based on their pharmacological properties and their final use, rather than their botanical source. Such a transition would require significant updates to international frameworks and domestic legislation, but it may be necessary to address the rapidly evolving landscape of cannabinoid science and commerce, offering consumers and governments a better understanding of the need, uses, and benefits of those cannabinoids as ingredients of products.

The path forward: toward scientifically grounded definitions

The current fragmentation of cannabis definitions represents a significant obstacle to rational policy development and international trade. As more countries move toward legalization, the need for harmonization becomes increasingly urgent. Organizations such as the ISO (International Organization for Standardization) have begun developing technical standards for the industry, while the UN's 2020 reclassification of cannabis represents a first step toward modernizing international frameworks [10].

For Mexico, adopting definitions aligned with emerging global standards while considering domestic priorities represents a strategic opportunity. Clear, scientifically grounded definitions would allow the country to leverage its comparative advantages in agriculture, manufacturing, and geographic location to capture value in the emerging global cannabis market. However, these technical choices must be made with full awareness of their broader social, economic, and cultural implications – the subject we turn to next in examining how historical stigma continues to shape policy implementation even after legal reform.

How definitions and other public health policies and stigmas can affect policy

Through the analysis of global markets, it becomes evident that legal definitions and conceptual frameworks are not mere technicalities but determining factors that profoundly shape the development of cannabis markets. Restrictive or confusing definitions have created artificial barriers to industry development in multiple jurisdictions, while more coherent and comprehensive approaches have facilitated the orderly growth of the sector.

In México and Brazil, for example, the rigid separation between "cannabic medicines" (allowed) and "cannabis" (prohibited) has created a two-tier system where wealthy patients legally access imported products while traditional users face criminalization [11]. In contrast, Canada's approach, which recognizes traditional uses alongside modern medical applications, has allowed for greater social and economic inclusion [12].

Public health policies also vary significantly in their approach to cannabis. While some countries have adopted a harm reduction approach, providing education on responsible consumption and promoting lower-risk products, others maintain prohibitionist messages that can, paradoxically, increase risk behaviors by limiting access to accurate information. However, these regulatory frameworks operate within broader social contexts where historical stigma continues to shape implementation outcomes.

Social stigma

Social stigma remains a determining factor even in legalized markets. In Australia, studies have documented how medicinal cannabis patients face discrimination in work and healthcare settings despite the legality of their treatments [13]. This residual stigma particularly affects vulnerable populations and can perpetuate inequalities in access even after legal reforms.

The financial impact of stigma manifests through discriminatory tax policies, banking restrictions, and inflated compliance costs. In 2025, a recent study conducted by Leafwell, published in the journal *Applied Health Economics and Health Policy*, suggests that the federal legalization of medical cannabis could lead to significant reductions in healthcare expenditures across the United States, potentially saving up to $29 billion per year [14].

Having worked with cannabis businesses across different regulatory frameworks, I've witnessed how the stigma translates into economic, banking, and financial barriers that would be considered unacceptable in any other industry. For instance, legal cannabis operations often face effective tax rates three to four times higher than comparable non-cannabis businesses due to tax codes rooted in the drug war era, representing a direct financial penalty for legalization [15].

During my most recent visit to Canada, I explored legal cultivation operations that evolved from legacy growers. The master grower, who serves as both owner and CEO of one such company, explained in detail the crushing financial burden they face after 7 years of legalization: between 42% and 70% of their revenue goes toward federal taxes, local taxes, storefront expenses, and licensing fees.

This creates a backwards payment structure where growers are the last in the supply chain to receive compensation. Here's how it works: they deliver their flower to the municipality, which then collects profits from the cannabis stores, deducts all applicable taxes, and finally pays the growers – but only if any profit remains. As a result, many legacy cannabis growers operate perpetually in debt, running in the red with no real access to fair market conditions.

For Mexico, these observations underscore the importance of accompanying any legal reform not only with comprehensive strategies for public education, professional training, and revision of social policies to address deep-rooted stigma, but also with a full understanding of the traceability, operations, banking, taxes, and how the financial structure of cannabis companies will develop and operate. The definitions adopted in the regulatory framework will not only determine technical aspects of the industry but will also communicate social values and set the tone for the cultural normalization of cannabis, and will determine the operability of the businesses involved.

Stigma and misconceptions around cannabis

Stigma and misconceptions around cannabis continue to shape how the plant, its users, and even emerging cannabis industries are perceived, often outweighing scientific evidence and lived experience. These narratives frame cannabis as inherently dangerous or morally suspect, reinforcing fear, discrimination, and resistance to reform even when medical or regulatory advances are made.

Cultural stigmas: origins of cannabis stigma and its ongoing impact

The stigma surrounding cannabis has deep and complex roots that intertwine with historical, political, and cultural factors. In Mexico, this stigma dates back to the early twentieth century, when the plant began to be associated with marginalized groups and "deviant" behaviors. The influence of the American "Reefer Madness" campaign spread internationally, labeling cannabis as a "drug of madmen" that supposedly caused violence, moral degeneration, and insanity.

This stigma was not accidental but the result of deliberate campaigns with both political and economic motivations. In the United States, Harry Anslinger, the first commissioner of the Federal Bureau of Narcotics, explicitly linked cannabis with racial and ethnic minorities, particularly Mexicans and African Americans, using xenophobic fears to drive prohibition [16]. Historical documents now reveal that Nixon's War on Drugs was explicitly designed as a political tool. John Ehrlichman, Nixon's domestic policy advisor, later admitted: "We knew we couldn't make it illegal to be either against the war or black, but by getting the public to associate the hippies with marijuana and blacks with heroin, and then criminalizing both heavily, we could disrupt those communities" [17].

Parallel to these racial and political motivations, industrial interests such as timber, paper, and synthetic fibers saw hemp as a potential competitor, benefiting from its prohibition. The power of these economic interests cannot be overstated. Newspaper magnate William Randolph Hearst, who owned vast timber holdings, used his media empire to spread anti-cannabis propaganda, while DuPont, developing synthetic fibers that would compete with hemp, actively lobbied for prohibition [18]. This economic motivation has been systematically obscured in public discourse about cannabis prohibition.

In Mexico, the criminalization of cannabis reinforced pre-existing social divisions, primarily stigmatizing rural and indigenous communities where cultivation and traditional use of the plant was more common. The derogatory term "mariguano" became a widespread insult, associated not only with consumption of the plant but with an entire series of negative characteristics: laziness, criminality, immorality, and lack of ambition.

This stigma has had devastating and lasting consequences. At an individual level, it has generated discrimination in employment, housing, and access to medical services. At a collective level, it has justified disproportionate punitive policies that have devastated entire communities. At a scientific level, it hindered legitimate research on the therapeutic potentials of the plant for decades, delaying important medical advances.

While stigma remains a significant consideration, changing perceptions demonstrate substantial momentum toward normalization and create opportunities for strategic intervention. Recent surveys showing approximately half the population supporting cannabis reform represent remarkable progress from the historical prohibitionist consensus, indicating that Mexican society is actively reconsidering cannabis policy. This evolving landscape provides multiple entry points for evidence-based education, professional training, and policy advocacy. International experience demonstrates that well-designed public education campaigns, professional medical training, and transparent regulatory implementation can accelerate stigma reduction while building public confidence in regulated cannabis systems.

Public perception versus science: how myths are dispelled through education and research

The discrepancy between public perception of cannabis and current scientific evidence represents one of the most significant challenges for rational policy formulation. For decades, myths and misinformation have dominated public discourse, creating a considerable gap between what science has discovered and what much of society, including professionals and decision-makers, believes about the plant.

Among the most persistent myths is the notion of cannabis as a "gateway drug" that inevitably leads to the use of more dangerous substances. Contemporary research has dismantled this theory, demonstrating that patterns of progression in substance use respond more to social, economic, and availability factors than to the pharmacological properties or cannabinoid percentages of cannabis itself [19].

Despite this outdated stigma, Mexico is developing its cannabis industry through innovative legal strategies. One emerging model is the cannabis social club – associations of consumers who have obtained permits for personal cultivation and consumption. These operate as nonprofit organizations with membership structures, providing associates access to a safe, educational platform for responsible use.

Beyond consumption, wellness companies are exploring the therapeutic benefits of minor cannabinoids, incorporating them into dietary supplements, food products, beverages, and cosmetics. Meanwhile, the hemp industry plays a crucial role in Mexico's cannabis landscape. Hemp-derived products for bio-construction, paper manufacturing, bioplastics, and other industrial applications remain legally distinct from recreational and medical cannabis markets, operating without the same regulatory restrictions or social stigma.

Another prevalent misconception is the idea that cannabis lacks significant medicinal value. This notion, even codified in international conventions that classified cannabis as having no therapeutic utility, has been refuted by a growing body of scientific evidence. Currently, the therapeutic applications of cannabis are supported by research demonstrating its efficacy for conditions such as chronic pain, spasticity, chemotherapy-induced nausea, and certain epileptic disorders, among others [20].

The idea that cannabis invariably causes brain damage or permanent cognitive impairment has also been significantly nuanced by modern research. While intensive consumption during adolescence presents risks for neurological development [21], longitudinal studies have questioned many of the previously reported associations between moderate adult consumption and lasting cognitive damage [22].

Evidence-based education has proven to be a powerful tool for dispelling these myths. In countries like Canada and Uruguay, government informational campaigns that present balanced information about benefits and risks have contributed to normalizing conversations about cannabis. Similarly, the incorporation of education about cannabinoids in medical curricula has significantly improved attitudes and knowledge of health professionals.

In Mexico, where public education about cannabis has historically been dominated by prohibitionist messages, there is an urgent need to develop evidence-based educational programs that provide objective information about risks and benefits, promoting informed decisions at both individual and political levels.

Breaking the stigma: the role of public policy, businesses, and advocacy in changing societal views

The transformation of deeply rooted social perceptions requires coordinated efforts from multiple sectors. Public policies play a fundamental role in this process, not only for their practical effects but also for their symbolic and normative function. When governments transition from criminalizing language to one centered on public health and rights, they send powerful signals that legitimize new ways of understanding cannabis.

Companies in the sector also have a significant responsibility in social normalization. Brands and companies that adopt professional, transparent, and ethically responsible approaches contribute to redefining the image of cannabis, distancing it from negative stereotypes. In regulated markets like Canada, companies that prioritize consumer education, regulatory compliance, and social responsibility have notably contributed to changing public perceptions.

Civil society organizations and advocacy groups have historically been crucial engines of change. In Mexico, organizations such as the RIA Institute, Mexican Cannabis Movement, Mexico United Against Crime, and patient groups have worked tirelessly to make visible both the harms of prohibition and the potential benefits of regulation. Their work has been instrumental in humanizing cannabis consumers and destigmatizing medical patients, providing a better understanding of drug policies and damage control.

The media exert a determining influence on the formation of public opinion. The evolution from sensationalist coverage to more nuanced and scientifically informed reporting has significantly contributed to destigmatization in countries where reform has advanced. In Mexico, although stereotypical representations persist, specialized media and committed journalists have begun to offer more balanced narratives.

Beyond formal education and policy reform, broader cultural narratives play an equally important role in shaping public perception. The cultural representation of cannabis in art, literature, and entertainment has a considerable impact on public perceptions. The emergence of cultural content that presents responsible uses and normalized contexts of cannabis, away from stigmatizing caricatures, gradually contributes to transforming the collective imagination.

For Mexico, a comprehensive approach to breaking stigma would require coordination between these various actors, combining legislative changes with educational campaigns, responsible business practices, effective advocacy, and normalized cultural representations. Only through this multidimensional approach can a stigma built and reinforced for almost a century be overcome.

Global cannabis landscape: how a better understanding supports better legal frameworks and markets

The transformation of deeply rooted social perceptions around cannabis represents one of the most significant challenges facing contemporary drug policy reform. While stigma remains a determining factor even in legalized markets, changing perceptions demonstrate substantial momentum toward normalization and create opportunities for strategic intervention. The evidence from countries that have advanced cannabis reform shows that well-designed public education campaigns, professional medical training, and transparent regulatory implementation can accelerate stigma reduction while building public confidence in regulated cannabis systems.

The development of internationally harmonized definitions will constitute a crucial step to facilitate an orderly global market. The current multiplicity of definitions of "cannabis," "hemp," "derivatives," and other fundamental terms creates unnecessary trade barriers, complicates transnational scientific research, and generates regulatory confusion. Organizations such as the ISO have begun to work on technical and more specific standards for the industry, an effort that deserves expanded support.

The primary insight gained both from the Mexican experience and the global landscape is that effective cannabis regulation is more than just a technical undertaking; it represents a transformative social process that requires a deep reconsideration of our cultural, medical, economic, and legal relationship with a plant that has accompanied humanity for millennia. The definitions and regulatory structures we build today will determine not only the development of an industry but also the future of drug policy, social justice, and access to therapeutic options for millions of people.

References

[1]   Shirk, D.A. (2011). *The Drug War in Mexico: Confronting a Shared Threat*. Council on Foreign Relations Special Report No. 60.

[2]   Ruiz, E. A. (2025). Medical Cannabis in Mexico: A Look at the Bill for Its Regulation. *Saúde & Sociedade*, SciELO 34(1), e240251pt. Accessed 24 Nov. 25, https://www.scielosp.org/article/sausoc/2025.v34n1/e240251pt/en/.

[3]   Holmes, O.W. Jr (1918). *Towne v. Eisner*, 245 U.S. 418. 425.

[4]   Santamarina y Steta (2023). *La regulación del Cannabis en México: S+S Insights*. Mexico City: Santamarina y Steta – Mariana Larrea Arias.

[5]   Código Sanitario de los Estados Unidos Mexicanos (2023). *Clasificación de Sustancias Psicoactivas y Psicotrópicas*. Mexico City: Secretaría de Salud.

[6]   United States Congress (2018). *Agriculture Improvement Act of 2018 (Farm Bill)*. Public Law. 115–334.

[7] Canada. Health Canada. *"Hemp and the Hemp Industry: Frequently Asked Questions."* Canada.ca, 12 Mar. 2025, https://www.canada.ca/en/health-canada/services/drugs-medication/cannabis/produc ing-selling-hemp/about-hemp-canada-hemp-industry/frequently-asked-questions.html. Accessed 25 Nov. 2025.

[8] European Union Drugs Agency. *"Cannabis – The Current Situation in Europe."* *European Drug Report 2025*, 5 June 2025, https://www.euda.europa.eu/publications/european-drug-report/2025/cannabis_ en. Accessed 25 Nov. 2025.

[9] Asociación Mexicana de Medicina Cannábica (2024) *Debate sobre Terminología Médica y Regulatoria del Cannabis*. Revista Mexicana de Cannabis Medicinal, 8(2), 45–52.

[10] United Nations Commission on Narcotic Drugs (2020). *Decision 63/12*: *Scheduling recommendations for cannabis and cannabis-related substances*. Vienna: UN Office on Drugs and Crime.

[11] Pinto CDBS, Esher Â, Oliveira CVDS, Osorio-de-castro CGS. (2024). Expansion of the medical cannabis market in Brazil and regulatory challenges. Cad Saude Publica, 40(11), e00088624. Published 2024 Dec 20. doi: 10.1590/0102-311XEN088624.

[12] Ontario Cannabis Store, and Deloitte Canada. *Six Years of Legalization: The Economic and Social Impact of Canada's Cannabis Sector*. August 2025. Ontario Cannabis Store / Deloitte Canada, https://www. doingbusinesswithocs.ca/wp-content/uploads/2025/08/OCS_Six-years-of-legalization-The-economic- and-social-impacts-of-Canadas-cannabis-sector_ENG-AODA.pdf. Accessed 24 Nov. 2025.

[13] Erku, D., Greenwood, L. M., Graham, M., Hallinan, C. M., Bartschi, J. G., Renaud, E., & Scuffham, P. (2022). From growers to patients: Multi-stakeholder views on the use of, and access to medicinal cannabis in Australia. *PloS one*, 17(11), e0277355. https://doi.org/10.1371/journal.pone.0277355.

[14] Chin, June, Do. (3 Oct. 2025). Medical Cannabis Programs: A Win for Businesses. *Leafwell Blog*, Accessed 24 Nov. 2025 https://leafwell.com/blog/cannabis-programs-business-win.

[15] Whitney, B. (2023). *The Tax Burden on Legal Cannabis: Quantifying the Stigma Premium*. Cannabis Business Executive, 15(4), 22–28.

[16] Anslinger, H. & Cooper, C. (1937). *Marihuana: Assassin of Youth*. American Magazine.

[17] Baum, D. (2016, April). *Legalize It All: How to win the war on drugs. Harper's Magazine*. https://harpers. org/archive/2016/04/legalize-it-all/

[18] McWilliams, J.C. (1990). *The Protectors: Harry J. Anslinger and the Federal Bureau of Narcotics, 1930–1962*. Newark: University of Delaware Press.

[19] Waltermaurer, Eve, Gerald Benjamin, and Leah Mancini. (2017). *The Marijuana Gateway Fallacy*. Discussion Brief, 18, The Benjamin Center for Public Policy Initiatives at SUNY New Paltz, Summer, www.newpaltz.edu/media/the-benjamin-center/db_18_the_marijuana_gateway_fallacy.pdf. Accessed 25 Nov. 2025.

[20] National Academies of Sciences, Engineering, and Medicine; Health and Medicine; Division; Board on Population Health and Public Health Practice; Committee on the Health Effects of Marijuana: An Evidence Review and Research Agenda (2017 Jan 12). The Health Effects of Cannabis and Cannabinoids: The Current State of Evidence and Recommendations for Research. Washington (DC): National Academies Press (US) Available from. https://www.ncbi.nlm.nih.gov/books/NBK423845/ doi:10.17226/24625.

[21] Scheyer, Andrew F et al. (2023). Cannabis in Adolescence: Lasting Cognitive Alterations and Underlying Mechanisms. *Cannabis and cannabinoid research*, 8(1), 12–23. doi: 10.1089/can.2022.0183.

[22] Gowin JL, Ellingson JM, Karoly HC et al. (2025). Brain Function Outcomes of Recent and Lifetime Cannabis Use. JAMA Network Open, 8(1) e2457069. doi: 10.1001/jamanetworkopen.2024.57069.

Robert T. Hoban

Why is *Cannabis* regulated? Global markets, national laws, and the hemp-marijuana divide

Introduction

The *Cannabis* plant has been cultivated and consumed by humans for thousands of years, yet in the twenty-first century it has become one of the most contested commodities in the global economy [1]. At the heart of the debate lies a paradox: while cannabis prohibition has historically been justified on public health and criminal justice grounds, the modern era has witnessed an unprecedented wave of legalization, commercialization, and international trade. Nations around the world are grappling with the challenge of designing cannabis laws that balance economic opportunity, public safety, and social equity [2]. In this context, the United States occupies a particularly complicated position – both as a country with vast cannabis production and consumption, and as a nation constrained by the federal divide between hemp and marijuana [3].

The term "cannabis" encompasses both hemp and marijuana. Botanically speaking, the two are the same species: *Cannabis sativa* L. What differentiates them under US law is the threshold of delta-9 tetrahydrocannabinol (THC), the psychoactive compound most associated with marijuana. The 2018 U.S. Farm Bill codified hemp as cannabis containing no more than 0.3% THC on a dry-weight basis [4]. Cannabis above this threshold is considered marijuana and remains a Schedule I controlled substance under federal law. This seemingly arbitrary distinction has generated enormous regulatory complexity. It has created thriving hemp markets in cannabinoids like cannabidiol (CBD), but has also subjected farmers, processors, and exporters to a patchwork of conflicting rules.

Meanwhile, the international marketplace has developed at a rapid pace. Canada legalized adult-use cannabis nationwide in 2018, becoming the first G7 nation to fully regulate production, distribution, and consumption [5]. Germany, Europe's largest economy, is rolling out a phased legalization approach, initially emphasizing medical cannabis and later experimenting with controlled recreational access [6]. Across Latin America and the Caribbean, countries such as Colombia, Uruguay, Jamaica, and Mexico have designed export-oriented regulatory systems aimed at supplying the growing global demand for medical cannabis [7]. Africa, too, has entered the market: Lesotho, Morocco, and South Africa are positioning themselves as low-cost producers with favorable climates [8].

As more nations embrace cannabis reform, international trade has become a central focus. The United Nations Single Convention on Narcotic Drugs (1961) still pro-

© 2026 Walter de Gruyter GmbH, Berlin | https://doi.org/10.1515/9783111476162-007

vides the backbone of global drug control, classifying cannabis as a controlled substance. Yet reinterpretations of this treaty, particularly following the 2020 vote by the UN Commission on Narcotic Drugs to reschedule cannabis for medical purposes, have opened the door to more robust cross-border commerce [9]. Governments are now confronted with questions of tariffs, quality standards, intellectual property protections, and supply chain integrity. For many emerging markets, cannabis represents both an agricultural development strategy and a means to attract foreign direct investment.

The United States however, remains constrained by its own contradictions. While more than 20 states have legalized adult-use marijuana and over 35 permit medical cannabis, federal prohibition remains firmly in place. The hemp carve-out of 2018 has created its own challenges, from the rise of intoxicating hemp derivatives like delta-8 THC to enforcement gaps that undermine both consumer safety and international credibility. This has placed US businesses in an awkward position: they are global pioneers in cultivation techniques, retail models, and branding, but cannot fully participate in the international marketplace as long as marijuana remains federally illegal. At the same time, US hemp companies have struggled with oversupply, inconsistent regulation, and difficulties accessing foreign markets that question the legitimacy of US hemp standards [10].

This chapter examines the dynamics of the emerging international cannabis marketplace, focusing especially on the US dilemma of hemp versus marijuana. It will explore how different nations are positioning themselves in the cannabis economy, the implications of the US legal divide, and the potential pathways toward a more coherent global cannabis framework. The chapter is organized into several sections: first, a historical overview of international cannabis trade and regulation; second, an analysis of the current state of the global marketplace; third, a deep dive into the US hemp-marijuana divide; fourth, case studies of key international players; and finally, a discussion of future challenges and opportunities in shaping a sustainable, equitable, and profitable cannabis industry.

The stakes are significant. By some estimates, the global legal cannabis market could exceed $100 billion before 2030, with ancillary industries adding even greater economic impact [11]. For developing countries, cannabis offers an opportunity to diversify economies, create jobs, and generate export revenues. For developed nations, cannabis is both a lucrative commercial sector and a policy arena tied to public health, taxation, and criminal justice reform. The international community must reconcile these forces with long-standing treaties and domestic politics. For the United States, resolving the hemp-versus-marijuana conundrum is not simply a domestic legal matter; it is a question of whether the country will lead or lag in the global cannabis economy.

Historical background: *Cannabis* and international regulation

Cannabis has long held a dual identity: as a utilitarian crop and as a psychoactive plant entangled in cultural, political, and legal controversy. Understanding the modern international cannabis marketplace – and the particular challenges for the United States – requires tracing the evolution of cannabis regulation, from ancient uses and colonial trade to twentieth-century prohibition and international treaties that continue to shape policy today.

Ancient and early uses of *Cannabis*

Archaeological evidence suggests cannabis has been cultivated for at least 10,000 years [1]. In ancient China, hemp was used for fiber, food, and medicine as early as 2700 BCE, with seeds recognized for nutrition and leaves and flowers for therapeutic purposes [12]. Hemp fiber became essential for rope, cloth, and paper, contributing to early technological advances. *Cannabis* spread westward through trade routes, reaching India – where it held religious significance as bhang – and the Middle East, where hashish appeared by the twelfth century. By the early modern period, cannabis was widely used for both industrial and psychoactive purposes.

Hemp and colonial expansion

Hemp's industrial value intensified during European colonial expansion [13]. Strong ropes and sails were critical to naval power, prompting colonies to cultivate hemp extensively. In North America, hemp was a staple crop in colonies like Virginia and Kentucky, with leaders such as George Washington and Thomas Jefferson documenting its cultivation. By the nineteenth century, cannabis extracts were common in Western medicine, appearing in pharmacies as treatments for pain, nausea, and insomnia. Hemp remained essential for textiles, ropes, and paper, reflecting the plant's dual-use nature until societal perceptions shifted in the early twentieth century.

The rise of prohibition

Modern cannabis prohibition emerged in the early 1900s, influenced in the United States by anti-immigrant sentiment. Mexican laborers introduced recreational smoking, which sensationalist media associated with crime and social deviance. The Marihuana Tax Act of 1937 effectively criminalized cannabis, sweeping hemp into prohibition and decimating its industry. Internationally, the League of Nations and subsequent anti-narcotics efforts increasingly treated cannabis as a dangerous substance alongside opium and cocaine [14].

The United Nations and global prohibition

The 1961 Single Convention on Narcotic Drugs codified global cannabis prohibition, placing it in Schedule IV, reserved for substances with little therapeutic value and high abuse potential [15]. Subsequent treaties – the 1971 Convention on Psychotropic Substances and the 1988 UN Convention Against Illicit Traffic – created a unified framework, forcing nearly every nation to limit cannabis use to medical and scientific purposes. Despite this, illicit cannabis markets flourished, with countries like Morocco, Afghanistan, Mexico, and Colombia becoming major suppliers.

Hemp's persistence

Despite prohibition, hemp endured for industrial purposes. France, Eastern Europe, and China maintained production for textiles, fiber, and seeds. By the late twentieth century, distinctions between hemp and marijuana reemerged, highlighting hemp's non-intoxicating nature. The EU subsidized hemp cultivation in the 1990s, and Canada legalized industrial hemp in 1998, paving the way for broader reforms.

Shifting attitudes and the twenty-first century

By the late twentieth century, cannabis attitudes began to shift. California's 1996 Proposition 215 legalized medical cannabis, inspiring other states and challenging federal prohibition. Globally, the Netherlands allowed limited recreational use, and Israel advanced therapeutic research. Hemp advocates emphasized sustainability, while economic pressures and scientific discoveries set the stage for reform. In the 2010s, Uruguay and Canada legalized nationwide, and US states expanded legal programs despite federal restrictions.

This historical trajectory – spanning ancient use, colonial expansion, prohibition, and gradual reform – frames the contemporary cannabis landscape. Understanding it is crucial to grasping the ongoing hemp-versus-marijuana divide in the United States and the complex dynamics of international cannabis regulation.

The global *Cannabis* marketplace today

The modern cannabis marketplace is no longer confined to underground economies or niche subcultures. Over the past two decades, cannabis has emerged as a legitimate agricultural, medical, and commercial product, valued by analysts at more than US $215 billion after 2030 [17]. The industry is global in scope: Canada's pioneering national legalization, Germany's pharmaceutical-led approach, Latin America's export orientation, Africa's entry into international supply chains, and Asia's tentative but important steps illustrate how diverse the cannabis sector has become. Despite differences in cultural attitudes and regulatory frameworks, a unifying reality has emerged: cannabis is now an international commodity, reshaping markets and challenging governments to rethink long-standing laws.

Canada: the first G7 nation to legalize

Canada occupies a unique position in the global cannabis marketplace. In 2018, it became the first G7 country – and the second nation after Uruguay – to legalize adult-use cannabis nationwide. This bold policy move transformed Canada into a laboratory for global cannabis regulation, attracting investors, entrepreneurs, and policymakers from around the world eager to observe its successes and shortcomings.

Canadian companies quickly expanded beyond domestic borders. Licensed producers such as Canopy Growth, Aurora Cannabis, Tilray, and Cronos raised billions of dollars in public markets, much of it on the Toronto Stock Exchange, and invested aggressively in cultivation and export facilities. Canada became a leading supplier of medical cannabis to Europe, Australia, and Latin America [18]. Its federally legal status allowed Canadian firms to participate in international trade in ways US companies could not.

However, Canada's experiment has also revealed the challenges of building a national cannabis industry from the ground up. Overproduction, burdensome regulations, and slow retail rollout resulted in falling stock prices and industry consolidation. Many companies wrote off billions in losses, and investor enthusiasm cooled. Yet despite these growing pains, Canada remains a benchmark. Its insistence on pharmaceutical-grade production (Good Manufacturing Practice, or GMP) set a high bar for quality that has influenced markets worldwide.

Germany and the European Union: the rise of a continental hub

Europe has approached cannabis reform more cautiously than North America, but momentum is building with Germany at the center. Germany legalized medical cannabis in 2017, creating a system that reimburses patients through public health insurance [6].

Imports of medical cannabis soared, primarily from Canada, the Netherlands, and Portugal, establishing Germany as the largest medical cannabis market in Europe.

In 2023, Germany passed legislation to expand medical cannabis access and introduce controlled recreational programs. This phased approach reflects both political compromise and European Union treaty constraints. Germany's policy is likely to influence neighboring states; already, Luxembourg, Malta, and Switzerland have adopted or piloted recreational reforms, while Italy, Poland, and the Czech Republic are expanding medical programs.

The European Union's role is crucial. While drug policy remains largely national, the EU's emphasis on pharmaceutical standards has set the tone for the continent. GMP certification has become a prerequisite for exporters seeking to access the European medical market. This regulatory rigor makes Europe an attractive but demanding destination for cannabis companies worldwide.

Latin America and the Caribbean: export-oriented strategies

Latin America has emerged as a promising region for cannabis cultivation and export. Blessed with favorable climates, low labor costs, and agricultural expertise, several Latin American countries view cannabis as a vehicle for rural development and foreign direct investment [7].

Uruguay was the first country in the world to fully legalize cannabis, in 2013. Its model emphasized social control rather than commercial expansion: production is tightly regulated, sales are limited to pharmacies, and advertising is prohibited. While Uruguay's domestic market remains small, its symbolic role as a pioneer cannot be overstated.

Colombia has taken a more export-oriented approach. With equatorial sunlight and rich agricultural soils, Colombia offers some of the lowest production costs in the world. The government has issued hundreds of licenses for cultivation and processing, aiming to position Colombia as a global supplier of medical cannabis and extracts. Canadian and European companies have invested heavily in Colombian operations, betting on its long-term export potential.

Mexico, the world's second-largest cannabis consumer market after the United States, has moved toward legalization, though political gridlock has slowed implementation. Once enacted, Mexico's policy could reshape the North American cannabis landscape, given its proximity to the United States and its vast agricultural capacity.

In the Caribbean, Jamaica has leveraged its cultural association with cannabis to build both a medical export industry and a tourism draw. Licensed Rastafarian sacramental use and "ganja tours" highlight the blend of cultural heritage and economic strategy. Other Caribbean nations, including Saint Vincent and the Grenadines, are exploring cannabis cultivation for export.

Africa: emerging suppliers

Africa has rapidly entered the conversation as a cannabis supplier. Favorable climates, low costs, and a history of informal cultivation have positioned African nations as future exporters, particularly to Europe. Lesotho was the first African country to issue medical cannabis licenses in 2017, attracting foreign investment from Canadian and South African firms. Its high-altitude climate is ideal for cannabis cultivation, and the government has actively promoted the industry as a source of economic growth.

Morocco, historically one of the world's largest illicit producers of cannabis resin (hashish), has begun transitioning toward legality [19]. In 2021, the government approved legislation to regulate cannabis for medical, industrial, and cosmetic purposes. Given Morocco's established cultivation expertise, it has the potential to become a leading exporter of regulated products. South Africa has also moved toward reform, with court decisions decriminalizing personal use and government initiatives exploring medical cannabis exports. Zimbabwe and Uganda have followed suit with cultivation licenses aimed at international markets [20].

Asia: tentative steps amid conservatism

Asia remains the most conservative region regarding cannabis, but significant developments are underway. Thailand legalized medical cannabis in 2018 and shocked observers in 2022 by decriminalizing the plant. Although its policies remain in flux, Thailand has emerged as a Southeast Asian hub for cannabis tourism and research. China dominates the global hemp industry, producing more than half of the world's hemp fiber and seeds. While marijuana remains strictly prohibited, China's role in hemp ensures it will remain central to at least part of the global cannabis supply chain [21]. India has allowed limited hemp cultivation under its Narcotic Drugs and Psychotropic Substances Act, though broader reform remains politically sensitive. Other Asian nations, including South Korea and Japan, permit limited medical research but have not embraced commercial cannabis [20, 22].

Looking ahead: an interconnected marketplace

The global cannabis marketplace is no longer hypothetical. It is a reality shaped by diverse regulatory models, economic strategies, and cultural attitudes. Canada and Germany have established themselves as pioneers; Latin America and Africa are positioning as low-cost exporters; Asia is cautiously entering the field. At the same time, international trade remains fragmented by treaties, quality standards, and politics. This international patchwork highlights both opportunities and challenges. For investors, the market offers new frontiers. For governments, it demands careful balancing of economic potential and public health. For the United States, it raises the pressing question of how to engage – or risk being left behind.

The US dilemma: hemp versus marijuana

The United States stands as both the world's largest consumer market for cannabis and one of the most fragmented legal jurisdictions for it. This paradox has profound consequences for the country's ability to participate in the emerging global cannabis marketplace. At the heart of the American cannabis conundrum is the legal distinction between "hemp" and "marijuana." Though both derive from the same species – *Cannabis sativa* L. – the federal government has chosen to treat them as radically different plants. Hemp, defined in law as cannabis with less than 0.3% delta-9 tetrahydrocannabinol (THC) on a dry-weight basis, was legalized under the 2018 Agriculture Improvement Act, commonly called the Farm Bill. Marijuana, by contrast, remains a Schedule I controlled substance under the federal Controlled Substances Act, alongside heroin and LSD.

This artificial distinction has created a regulatory quagmire. On the one hand, hemp's legalization sparked a gold rush in cultivation, processing, and product development, especially around cannabidiol (CBD). On the other hand, marijuana's ongoing federal prohibition continues to hinder the ability of US companies to access capital, banking, insurance, and interstate or international trade. The result is a divided marketplace: hemp and hemp-derived cannabinoids circulate widely across state and national borders, while marijuana is trapped in a patchwork of state-level regimes.

The 2018 Farm Bill and its consequences

The 2018 Farm Bill was a watershed moment. For the first time since the Marihuana Tax Act of 1937, American farmers could legally cultivate hemp as an agricultural commodity. The law opened the door for hemp-derived cannabinoids, especially CBD, to enter mainstream consumer markets. Overnight, CBD appeared on pharmacy shelves, in wellness products, beverages, cosmetics, and even pet supplements.

Yet the Farm Bill's success was short-lived. Farmers, eager to capitalize on the "green rush," planted hundreds of thousands of acres of hemp. By 2019, supply vastly exceeded demand, sending prices crashing and leaving many farmers with unsellable crops. Regulatory uncertainty compounded the problem. The Food and Drug Administration (FDA) maintained that CBD could not legally be marketed as a dietary supplement or added to food, creating confusion for businesses and consumers alike. Enforcement was inconsistent, leading to a marketplace filled with products of widely varying quality and safety.

The Farm Bill also introduced the now-famous 0.3% THC threshold. Borrowed from Canadian and European models, this limit was originally intended as an administrative convenience rather than a scientifically meaningful line. In practice, it has proved arbitrary and problematic. Farmers often see their crops test just above the threshold, forcing them to destroy fields deemed "hot." Products derived from hemp with trace amounts of THC can quickly slip into a legal gray zone.

The rise of intoxicating hemp derivatives

Perhaps the most unintended consequence of the 2018 Farm Bill has been the rise of intoxicating hemp-derived cannabinoids. Because the law legalized hemp and its "derivatives, extracts, and cannabinoids," entrepreneurs quickly discovered that they could synthesize psychoactive compounds from hemp-derived CBD. Substances such as delta-8 THC, delta-10 THC, THC-O acetate, and HHC (hexahydrocannabinol) emerged as consumer products, often marketed as legal alternatives to marijuana.

These compounds exploit a loophole: they are chemically similar to delta-9 THC but not explicitly banned under federal law. Because they are derived from federally legal hemp, producers claim they fall within the Farm Bill's definition. The result has been an explosion of products that mimic marijuana intoxication but circulate through convenience stores, smoke shops, and online platforms outside the regulated cannabis dispensary system [23].

This development has alarmed regulators, health officials, and traditional cannabis businesses alike. Critics argue that intoxicating hemp derivatives undermine consumer safety, evade taxation, and blur the lines between hemp and marijuana even further [24]. Several states have moved to restrict or ban these products, but the lack of federal clarity leaves the issue unresolved.

Marijuana: state legalization versus federal prohibition

While hemp has been legalized, marijuana remains federally illegal. Yet at the state level, legalization has advanced rapidly. As of 2025, more than 20 states have legalized adult-use marijuana, and over 35 allow medical cannabis programs. These state regimes have created thriving industries, generating billions in tax revenue and employing hundreds of thousands of people.

But the lack of federal reform creates significant barriers. Cannabis businesses cannot ship products across state lines, forcing each state market to operate as a silo. Banking and financial services remain difficult to access, as many institutions fear federal penalties [25]. Interstate commerce, international exports, and even routine business activities like securing insurance or deducting expenses on federal taxes are hindered by marijuana's Schedule I status.

This disjointed system leaves the US cannabis industry both advanced and handicapped: advanced in its consumer sophistication, retail innovation, and cultivation techniques; handicapped in its inability to scale nationally or internationally.

International implications

The hemp-marijuana divide also undermines US credibility in the global marketplace. Canadian companies can export medical cannabis to Europe under federal authorization, while US firms cannot ship marijuana even between neighboring states. Latin American and African countries are positioning themselves as export hubs, while American businesses remain locked out of international trade.

Even within the hemp sector, challenges persist. Many foreign governments view US hemp regulation as lax, particularly given the prevalence of intoxicating hemp derivatives. This has complicated efforts to establish reliable export markets for US hemp products. Without harmonization between hemp and marijuana regulations, the United States risks falling behind as the global cannabis industry matures.

Challenges: a house divided

The US hemp-marijuana divide is more than a legal curiosity. It is a structural barrier that hampers domestic industry, undermines international competitiveness, and creates consumer confusion. Hemp was intended to be an agricultural commodity, but its regulation has spawned unintended markets and loopholes. Marijuana thrives at the state level, but federal prohibition prevents the industry from realizing its full potential.

Until the United States reconciles these two sides of the same plant, it will remain a paradox: a global leader in cannabis innovation, retail, and culture – but a laggard in international trade and regulatory coherence. The next phase of cannabis reform must address this dilemma head-on if the United States hopes to claim a central role in the international cannabis economy.

Pathways to harmonization

The rapid internationalization of the cannabis industry has created a paradox. On the one hand, cannabis is increasingly treated as a legitimate agricultural commodity, medical therapy, and consumer product. On the other hand, regulation remains fragmented, inconsistent, and often contradictory. The challenge of harmonization is therefore central to the future of the cannabis marketplace. Without alignment across borders, businesses face legal uncertainty, governments miss out on tax revenues, and consumers are left with uneven access and product safety.

Harmonization refers to the process of aligning standards, rules, and frameworks across jurisdictions to facilitate trade, ensure safety, and promote stability. In the context of cannabis, harmonization encompasses quality standards, international treaties, intellectual property protections, and the broader architecture of global governance. This section explores these pathways, emphasizing the opportunities and obstacles that define them.

Good Manufacturing Practice (GMP) standards

The single most important driver of harmonization in the cannabis sector to date has been the application of GMP standards [26]. Originating in the pharmaceutical industry, GMP requires rigorous controls over production, testing, and distribution to ensure product safety, quality, and consistency.

Europe, particularly Germany, has made GMP certification a prerequisite for medical cannabis imports. This has effectively established GMP as the global standard for cannabis destined for medical markets. Canadian, Colombian, Portuguese, and Israeli producers seeking to export; Europe must invest heavily in GMP compliance, creating a de facto global rulebook.

GMP standards serve several purposes: they protect patients, reassure regulators, and create a level playing field for international trade. At the same time, they raise barriers to entry. Small-scale farmers in Africa or Latin America often lack the resources to meet GMP requirements, limiting their participation in lucrative export markets. The challenge is how to extend harmonization without excluding traditional growers or creating monopolistic advantages for large corporations.

Phytosanitary rules and agricultural trade

Cannabis, as an agricultural product, is also subject to phytosanitary regulations designed to prevent the spread of pests and diseases across borders. These rules are enforced through the International Plant Protection Convention (IPPC) and the World Trade Organization's Sanitary and Phytosanitary (SPS) Agreement [27, 28].

Exporters must comply with testing, quarantine, and certification requirements to ship cannabis internationally. For example, the European Union requires strict screening for contaminants such as mold, heavy metals, and pesticide residues. Canada, Israel, and Colombia have invested in laboratories and compliance systems to meet these demands.

While phytosanitary rules are essential for agricultural safety, their uneven application creates trade frictions. Some countries enforce rigorous standards, while others lack the infrastructure for testing and certification. Harmonization in this area requires investment in laboratory capacity, the development of shared testing protocols, and international cooperation to avoid duplication or contradictory requirements.

Intellectual property and plant variety protection

Another critical aspect of harmonization is intellectual property (IP). As cannabis becomes a global commodity, questions arise over the ownership of genetics, breeding innovations, and proprietary cultivation methods [29]. The World Intellectual Property Organization (WIPO) and the International Union for the Protection of New Varieties of Plants (UPOV) provide frameworks for plant variety protection. These systems allow breeders to register unique strains or cultivars, giving them exclusive rights to commercialize those varieties. Companies in Canada, the Netherlands, and Israel have already begun registering cannabis varieties, while US firms file patents despite federal prohibition.

The challenge lies in balancing innovation with equity. Many cannabis genetics originate from indigenous and traditional farming communities in places like Morocco, India, and Jamaica. Granting exclusive rights to corporations without recognizing traditional knowledge raises concerns about biopiracy. Harmonization in IP therefore requires not only technical standards but also ethical frameworks that acknowledge and compensate source communities.

International treaties and the United Nations

Perhaps the greatest barrier to harmonization lies in international law. The 1961 Single Convention on Narcotic Drugs, along with the 1971 and 1988 treaties, continues to classify cannabis as a controlled substance. These treaties obligate member states to limit cannabis to medical and scientific use, constraining the ability of countries to legalize recreational markets.

Yet reinterpretations and reforms are emerging. In 2020, the UN Commission on Narcotic Drugs voted to remove cannabis from Schedule IV, acknowledging its medical value. While cannabis remains in Schedule I, this shift has opened space for national reforms [9]. Uruguay, Canada, and Malta have moved forward with legalization despite treaty obligations, arguing that flexibility exists under the conventions. Germany and others have designed phased approaches to minimize treaty conflicts.

Future harmonization may require either amending the conventions or reinterpreting them in ways that accommodate legalization. Precedents exist: the international drug control system has adapted in the past to shifting realities. Regional cooperation – for example, within the European Union or the Organization of American States – may also provide a pathway to coordinated reforms that push the global system to evolve.

Financial services and trade infrastructure

Another dimension of harmonization is financial. Cannabis businesses face barriers to banking, insurance, and capital markets, particularly in the United States where federal prohibition persists [30]. Internationally, inconsistent rules on money laundering, taxation, and investment create uncertainty for cross-border finance. Harmonization here would mean creating clear standards for financial compliance, tax reporting, and trade financing. Organizations such as the Financial Action Task Force (FATF), which sets global standards on money laundering, could play a role. Similarly, free trade agreements may eventually incorporate cannabis provisions, just as they do for alcohol and tobacco.

Safety and labeling

Harmonization is also essential for consumer protection. Without shared standards, products can vary widely in potency, purity, and labeling. This creates risks for patients and undermines confidence in the industry. Efforts are underway to create international labeling standards that specify cannabinoid content, terpene profiles, dosing instructions, and warnings. The International Organization for Standardization (ISO) has begun developing cannabis-specific guidelines, while groups such as ASTM International, the American Herbal Products Association (AHPA), and the American Herbal Pharmacopoeia (AHP) have issued voluntary standards and guidance documents [16]. If widely adopted, these could provide the basis for global harmonization.

Takeaways: toward a coherent global framework

The future of cannabis lies not only in national reforms but in the creation of a coherent international framework. GMP standards, phytosanitary rules, intellectual property protections, treaty reinterpretations, financial compliance, and consumer labeling all represent essential pieces of the puzzle. Harmonization will not mean uniformity; countries will continue to tailor policies to their own needs. But greater alignment is essential for unlocking the full potential of the global cannabis industry. Without it, trade will remain fragmented, innovation will be stifled, and consumers will face uncertainty. With it, cannabis can move from the margins to the mainstream of international commerce, benefiting farmers, patients, businesses, and governments alike.

Future outlook: geopolitical and economic implications

The global cannabis industry has advanced from the margins of the underground economy into the mainstream of policy debates, investment portfolios, and agricultural development. Yet the future trajectory of cannabis remains uncertain. The industry is shaped not only by consumer demand and business innovation but also by geopolitics, international trade law, public health concerns, and evolving cultural attitudes. The United States – with its contradictory hemp and marijuana frameworks – will continue to play a pivotal role in determining whether the global cannabis economy becomes a fragmented patchwork or a harmonized marketplace.

This next section examines key scenarios for the industry's future, exploring the geopolitical and economic implications of reform, the role of multilateral institutions, the potential for US leadership or marginalization, and the broader impact of cannabis on global commerce.

US federal reform scenarios

The United States is the largest cannabis consumer market in the world, yet its contradictory legal framework continues to prevent it from fully engaging in the international marketplace. Over the next decade, several possible scenarios could unfold [3]. One path is descheduling or rescheduling cannabis under the Controlled Substances Act. Removing cannabis from Schedule I entirely would enable it to be regulated like alcohol or tobacco, opening the door to interstate commerce, banking, and global trade. Rescheduling to Schedule II or III would not go as far but would still ease restrictions, encourage medical research, and strengthen the US credibility abroad. Unlike Schedule II, Schedule III would remove banking and finance restrictions. Another path is incremental federalism, in which Congress passes measures such as the SAFE Banking Act or the STATES Act to address narrow issues like financial access and state-level autonomy, without fully descheduling cannabis [31]. Finally, continued gridlock remains a possibility, where federal inaction forces states to expand legalization independently while prohibition persists at the national level.

The trajectory the United States ultimately chooses will have profound consequences for the global cannabis economy. Federal descheduling would likely accelerate international reforms and establish new standards for trade, positioning the United States as a leader in shaping the industry. By contrast, continued prohibition would leave the United States isolated and unable to capitalize on the market it helped pioneer, as countries like Canada, Germany, and Colombia capture greater influence and market share.

Geopolitical dynamics

Cannabis is increasingly entangled in geopolitics, as countries seek to position themselves for both domestic consumption and international export dominance. In North America, Canada has leveraged its federal framework to expand internationally, becoming a global pioneer in cannabis exports. If the United States were to deschedule cannabis at the federal level, North America could emerge as the undisputed hub of the global cannabis industry, combining Canada's early mover advantage with the sheer size of the US consumer market. Mexico's progress toward legalization further strengthens this possibility, as it could create a fully integrated North American cannabis economy.

In Europe, Germany's leadership has proven decisive. Its medical-first approach, combined with the European Union's strict regulatory standards, has made Europe the global benchmark for quality and compliance. Other European nations are already following Germany's cautious but deliberate model, laying the foundation for a continental framework that exporters worldwide must respect if they wish to access this highly lucrative market. This emphasis on GMP standards is setting the tone for how cannabis is grown, processed, and distributed well beyond Europe's borders.

Meanwhile, emerging regions are carving out their own strategies. Latin American countries such as Colombia and Uruguay view cannabis as a tool for economic development, aiming to become the world's low-cost suppliers much as they have in industries like coffee and cut flowers. Africa's entry points – Morocco, Lesotho, and South Africa – reveal similar ambitions, but success will depend on moving beyond raw cultivation to include processing and branding. In Asia, most countries remain conservative, yet Thailand's reforms and China's dominance in hemp ensure the region will play a role, while India holds the potential to reshape the entire Asian cannabis landscape if it embraces reform.

Conclusion: the next decade

The next decade will be decisive for cannabis. Internationally, more countries will move from prohibition to regulation, driven by economic incentives, patient demand, and shifting cultural norms. Multilateral institutions will be forced to adapt, whether through reinterpretation or reform of treaties. Harmonization of standards will gradually reduce fragmentation, enabling more robust global trade.

The United States faces a choice: remain hamstrung by outdated laws or embrace reform and shape the global marketplace. For investors, entrepreneurs, and policymakers, the opportunity is immense. For farmers and patients, the stakes are personal. For governments, cannabis represents both risk and reward. The global cannabis economy is not just emerging – it is inevitable. The question is not whether cannabis will become a normalized part of international trade, but how quickly, under what rules, and who will lead in defining its future.

References

[1] Abel EL. (1980).Marihuana, The First Twelve Thousand Years. Plenum Press. (Plenum Press).

[2] Hall W, Stjepanović D, Caulkins J, Lynskey M, Leung J, Campbell G et al. (2019). Public health implications of legalising the production and sale of cannabis for medicinal and recreational use. Lancet, 394(10208), 1580–1590.

[3] Marcu JP, Schechter JB. (2024). The role of FDA in defining Cannabis and hemp: A brief perspective on the future of Cannabis regulation in the United States. Cannabis Innovations: Regulated Cannabis and Hemp Market Navigation, Business, Public Health, and Research Challenges, 15–40.

[4] The 2018 Farm Bill's Hemp Definition and Legal Challenges to State Laws Restricting Certain THC Products | Congress.gov | Library of Congress [Internet]. [cited 2025 Oct 12]. Available from: https://www.congress.gov/crs-product/R48637

[5] Belackova V, Petruzelka B, Cihak J, Michailidu J, Mravcik V. (2025). Getting "The whole picture": A review of international research on the outcomes of regulated cannabis supply. International Journal of Drug Policy, 142, 104796.

[6] Hofmann R. (2023). The 'Total-Legalization' of Cannabis in Germany: Legal Challenges and the EU Free Market Conundrum. European Journal of Crime, Criminal Law and Criminal Justice, 31(2), 173–196.

[7] Carrillo S. (2024).Cannabis innovations and regulation in Latin America. Regulated Cannabis and Hemp Market Navigation: Business, Public Health, and Research Challenges. Jahan Marcu and Andrew M. Peterson (edited by). Berlin, Boston:De Gruyter. Marcu [Jahan Peterson] Andrew M., (eds) 41–55. Available from: https://doi.org/10.1515/9783110677508-003.

[8] Anywar G, Kakudidi E, Tugume P, Asiimwe S. (2022). The Cannabis/Marijuana (Cannabis sativa L.) Landscape in Africa: an overview of its cultivation and legal aspects. Cannabis/Hemp for Sustainable Agriculture and Materials, 297–310.

[9] UN commission reclassifies cannabis, yet still considered harmful | UN News [Internet]. [cited 2025 Oct 12]. Available from: https://news.un.org/en/story/2020/12/1079132

[10] National Hemp Report. (February 2022). USDA, National Agricultural Statistics Service https://esmis. nal.usda.gov/sites/default/release-files/gf06h2430/xd07hw825/v692v917t/hempan22.pdf.

[11] Cannabis Market Size Projections: $100 Billion in the US by 2030 – Business Insider [Internet]. [cited 2025 Oct 12]. Available from: https://www.businessinsider.com/us-cannabis-market-size-projection -100-billion-by-2030-2020-12

[12] Russo EB, Jiang HE, Li X, Sutton A, Carboni A, Bianco F del et al. (2008). Phytochemical and genetic analyses of ancient cannabis from Central Asia. Journal of Experimental Botany, Oct 21. 59(15). 4171–4182.

[13] Schroeder M. The history of European hemp cultivation | Dissertations in Geology at Lund University [Internet]. [cited 2025 Oct 12]. Available from: https://lup.lub.lu.se/student-papers /search/publication/8985524

[14] Marcu J, Schechter JB. (2025). The Landscape of Cannabis Education. Cannabis and the Developing Brain. Marusak"] ["Hilary, editor; 597–622. Available from: https://doi.org/10.1007/978-3-031-87990-6_25

[15] Single Convention on Narcotic Drugs (1961). United Nations [Internet]. [cited 2025 Oct 12]. Available from: https://www.unodc.org/pdf/convention_1961_en.pdf

[16]   Russo EB. (2016). Current Therapeutic Cannabis Controversies and Clinical Trial Design Issues. Frontiers in Pharmacology, 7, 309.

[17]   Legal Marijuana Market Size, Share & Growth Report, (2030) [Internet]. [cited 2025 Oct 12]. Available from: https://www.grandviewresearch.com/industry-analysis/legal-marijuana-market

[18]   Case Studies In International Cannabis: What The World Can Teach Us [Internet]. [cited 2025 Oct 12]. Available from: https://www.forbes.com/sites/roberthoban/2025/09/05/case-studies-in-international-cannabis-what-the-world-can-teach-us/

[19]   Tinasti K, Afsahi K, Mouna K. (2025). Navigating cannabis regulation in Morocco: Historical context, socio-economic impact, and policy gaps. Journal of Illicit Economies and Development, 7(2), 53–60.

[20]   Howell S, Rusenga C. (2025). Comparative analysis of cannabis legalization in South Africa and Zimbabwe: Trajectories, commonalities, and divergences. Journal of Illicit Economies and Development, 7(2), 70–84.

[21]   McPartland JM, Hegman W, Long T. (2019). Cannabis in Asia: its center of origin and early cultivation, based on a synthesis of subfossil pollen and archaeobotanical studies. Vegetation History and Archaeobotany, 1–12.

[22]   Thippaiah SM, Gude JG. (2025). Recreational Use of Cannabis: Insights from India and Other Asian Countries. Journal of Psychiatry Spectrum, 4(3), 211–217.

[23]   Marcu JP, Simon TA. (2022). 5047 – In Flagrante Delicto Derivatives of Hemp-CBD. 2022 American Public Health Association Annual Meeting & Expo [Internet]. American Public Health Association Annual Meeting & Expo; [cited 2024 Aug 19]. Available from: https://apha.confex.com/apha/2022/meetingapp.cgi/Paper/518929

[24]   Simon TA, Simon Jh, Heaning EG, Gomez-Caminero A, Marcu JP. (2023). Delta-8, a Cannabis-derived tetrahydrocannabinol isomer: Evaluating case report data in the food and drug administration adverse event reporting system (FAERS) database. Drugs, Healthcare and Patient Safety, 15, 25–38.

[25]   Pennington S, Kline A, Mullin HB, Marcu J, Simon T. (2023). Coalition for Cannabis Scheduling Reform Report [Internet]. Available from: https://schedulingreform.org

[26]   21st Century Cannabis: Getting The World In Tune [Internet]. [cited 2025 Oct 12]. Available from: https://www.forbes.com/sites/roberthoban/2020/09/28/21st-century-cannabis-getting-the-world-in-tune/

[27]   IPPC – International Plant Protection Convention [Internet]. [cited 2025 Oct 12]. Available from: https://www.ippc.int/en/

[28]   WTO | Sanitary and Phytosanitary Measures – text of the agreement [Internet]. [cited 2025 Oct 12]. Available from: https://www.wto.org/english/tratop_e/sps_e/spsagr_e.htm

[29]   Fortin K, Charbonneau R. (2024). Intellectual property in the Cannabis industry. In: Cannabis Innovations: Business, Public Health, and Research Challenges [Internet]. De Gruyter. 105–117. Available from https://doi.org/10.1515/9783110677508-006.

[30]   Zender J, Eschker E, Gold G. (2025). Financial management in the Cannabis Industry: : Innovative financial practices in the cannabis sector. Strategic Finance, 1–7,

[31]   H.R.2891 – 118th Congress (2023–2024): SAFE Banking Act of 2023 | Congress.gov | Library of Congress [Internet]. [cited 2025 Oct 12]. Available from: https://www.congress.gov/bill/118th-congress/house-bill/2891

Ruth Fisher

One dose does not a user make: systemic failures in cannabis data collection

Introduction

Does cannabis cause schizophrenia, psychotic disorders, anxiety, depression, or other mental illness [1]? Heart attacks or strokes [2]? Lung cancer [1]? Some other health problem? There are tens of thousands of studies that have attempted to determine the impact of using cannabis on various adverse health effects. Most of these studies are retrospective in nature, that is, they compare outcomes for people who have used cannabis to those who have not, and they overwhelmingly conclude that cannabis users fared worse than non-users. Yet, the reported outcomes of a significant portion of these studies are inaccurate at best, often fatally so. In other words, a lot of published cannabis research is bad.

The question of whether or not cannabis users experience a greater prevalence of adverse health effects than their nonusing counterparts hinges on the ability of researchers to meaningfully and accurately distinguish cannabis users from non-users. Yet, this is exceedingly difficult to do, for two reasons. The first problem is the lack of a clear understanding among researchers and healthcare professionals as to what constitutes clinically meaningful cannabis use. That is, how much or how frequently must an individual use cannabis for it to meaningfully impact that person's health? The second problem is researchers' inability to accurately identify people who qualify as cannabis users. Unless or until researchers definitively establish what constitutes meaningful cannabis use, accurately identify these users from non-users in their studies, and then firmly establish whether or not such cannabis use is detrimental to health, a lot of bad cannabis research will continue to be published.

What constitutes clinically meaningful cannabis use?

The first problem impeding researchers' ability to accurately assess the potential health impact of cannabis use is a lack of clear understanding of the patterns in cannabis use that are clinically meaningful. The term "clinically meaningful" is an ambiguous term, subject to a wide variety of interpretations. While cannabis may generate both beneficial and/or adverse health effects in consumers, the present discussion focuses on studies of cannabis use correlated with adverse effects. Yet, all the general concepts discussed also extend to studies of cannabis use correlated with beneficial effects. For the present purposes, then, the question to be explored is, how much cannabis use is associated with adverse health effects?

Commonly cited health problems associated with cannabis use

There are a number of acute adverse side effects associated with cannabis use, including acute anxiety or panic attacks, GI problems, cognitive impairment, and acute cardiovascular changes. However, these tend to be temporary effects associated with single instances of use [3].

The most commonly cited serious health effects associated with more than a single dose of cannabis include:

- **Schizophrenia/psychoses**: "Cannabis use is likely to increase the risk of developing schizophrenia and other psychoses; the higher the use, the greater the risk" [4].
- **Impaired cognitive function**: "Persistent [over 20 years] cannabis use was associated with neuropsychological decline broadly across domains of functioning . . ." [5].
- **Respiratory problems**: "Cannabis smoking via combustion, even without cigarette smoking, has been associated with respiratory symptoms . . ." [6].
- **Anxiety/depression**: "Regular [daily or nearly every day] marijuana use is associated with an increased risk of anxiety, depression, and psychotic illness . . ." [7]. Notably, cannabis is known to exhibit biphasic effects, where lower doses of THC are associated with lower levels of anxiety and depression, while higher doses are associated with higher levels of anxiety and depression [4, 8].
- **Cannabis use disorder**: "There is substantial evidence for a statistical association between increases in cannabis use frequency and the progression to developing problem cannabis use" [4].
- **Reduced motivation**: "[T]here is evidence that long-term heavy [daily] cannabis use is associated with educational underachievement and impaired motivation . . ." [8].
- **Chronic cardiovascular problems**: "Cannabis use is associated with adverse cardiovascular outcomes [coronary heart disease, myocardial infarction, and stroke], with heavier [daily] use . . . associated with higher odds of adverse outcomes" [2].

It is currently believed that cannabis use does not cause schizophrenia or psychoses per se, but rather that using cannabis may hasten the onset of schizophrenia or psychoses in users who are predisposed toward such conditions [9, 10]. As for impaired cognitive function and reduced motivation, those problems tend to be most strongly associated with chronic cannabis use during adolescence [4, 8].

For the remaining most commonly cited associated serious health effects, the above descriptions suggest these problems tend to be associated with "persistent," "higher doses," "long-term heavy use," or "heavier use," generally defined within those studies as daily or near daily use. So if we define "meaningful cannabis use" to be use that is more likely to be correlated with these types of serious effects, then meaningful cannabis use can be said to involve daily or near daily use. In this case, an understanding of the size of the at-risk population requires understanding what portion of cannabis users are daily users.

Prevalence of cannabis use

Monitoring the Future (MTF), an organization funded by the National Institute on Drug Abuse (NIDA), started collecting information in 1975 on substance use in the population. MTF captures information on the prevalence of cannabis use within the past day, month, and year [11] (Figure 1). The MTF data indicate that the prevalence of cannabis use in the adult population has been increasing over time, and it is generally highest among 19–30-year-olds, with prevalence of use decreasing with age thereafter.

As of 2023, about 42% of 19–30-year-olds and about 29% of 35–50-year-olds reported having used cannabis within the past 12 months. About 29% of 19–30-year-olds and 19% of 35–50-year-olds have used it within the past 30 days. And about 10% of 19–30-year-olds and 7.5% of 35–50-year-olds reported daily use (Figure 1). Equivalently, about 25% of those reporting cannabis use in the past 12 months reported daily use (Figure 2). Reported 12 month, 30 day, and daily use by 55–65-year-olds (not shown in the figures) was, respectively, 19%, 14%, and 5%.

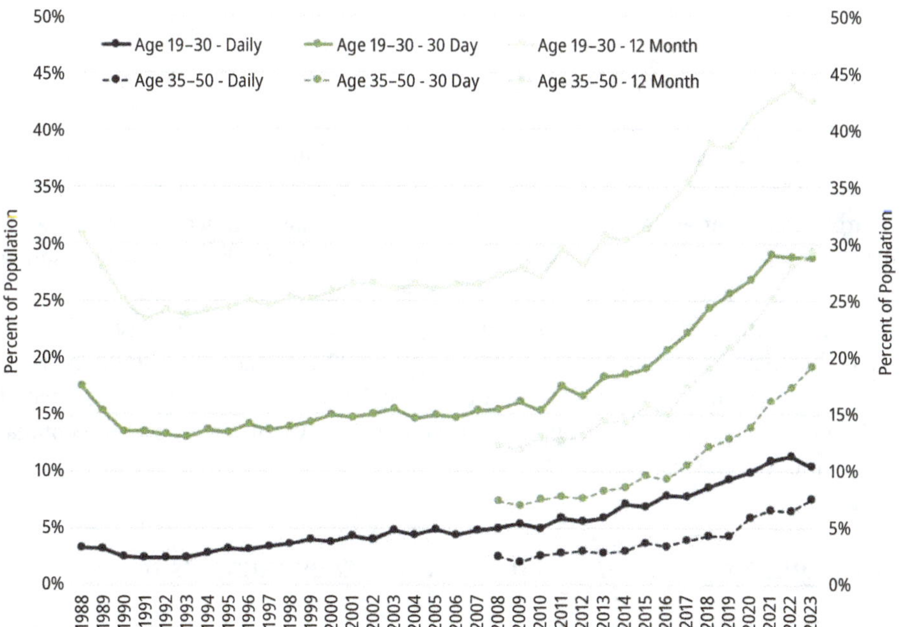

Figure 1: Estimated daily, 30-day, and 12-month prevalence of cannabis use by adults. Source: Patrick et al. [11].

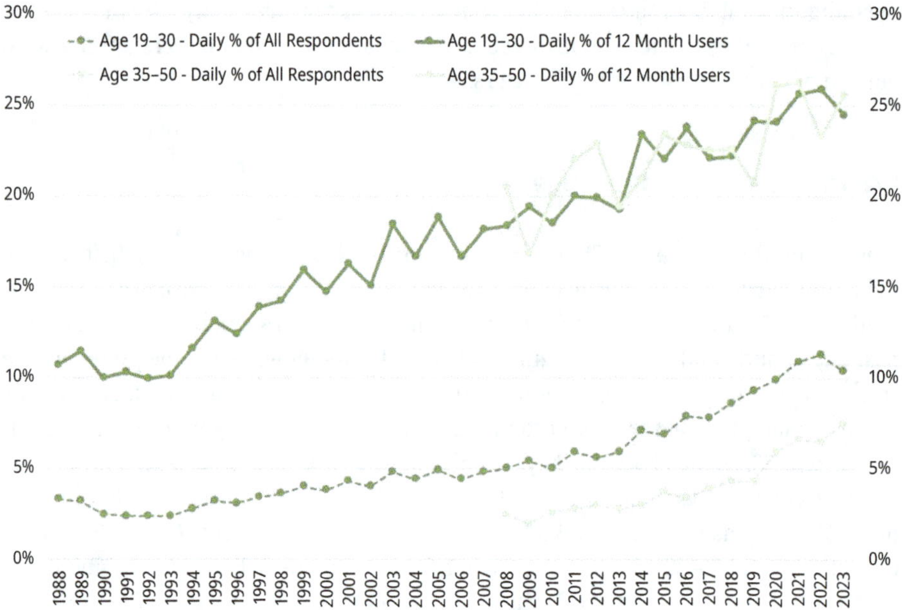

Figure 2: Estimated prevalence of daily cannabis use and daily cannabis use as a percent of 12-month cannabis use.
Source: Patrick et al. [11].

Prevalence of cannabis abuse

It is currently recognized that cannabis has a potential for abuse. The Diagnostic and Statistical Manual of Mental Disorders, DSM-5, defines cannabis use disorder (CUD) as "a problematic pattern of cannabis use leading to clinically significant impairment or distress, as manifested by at least 2 of 11 stated criteria occurring within 12 months" [12]. The US Centers for Disease, Control and Prevention (CDC) indicates that "recent [2021] research estimated that approximately 3 in 10 [30%] people who use marijuana have marijuana use disorder" [13]. That is, about 30% of past year cannabis users have been diagnosed as having CUD, though the CDC's estimate of the prevalence of CUD among cannabis users likely overstates the true prevalence of CUD in the population, for a variety of reasons [14, 15].

Distinguishing meaningful users from other cannabis users

From the earlier discussion, the MTF data suggest that as of 2023, roughly 30% of the adult population (age 19–65) reported use within the past year, about 20% reported use within the past month, and about 8% reported daily use. Otherwise stated, about a quarter of adults who used cannabis within the past year used cannabis daily. Taken as a whole, the information thus suggests that about a quarter of past-year cannabis users are meaningful users, that is, more prone to develop a serious adverse health effect associated with chronic cannabis use.

An important point of note is that a large portion of the population can be described as cannabis users; almost a third of the population reported using cannabis over the past year, and about a fifth reported using cannabis within the past month. Conversely, a relatively small portion of the population, less than a tenth, reported using cannabis daily, and it is this group of individuals that is more susceptible to experiencing serious health effects associated with that use. At the same time, a significant portion of this at-risk group, perhaps a third to a half [4, 16], uses cannabis medically. Relative to nonmedical users, medical cannabis users are less likely to suffer from cannabis-induced problems, due to both a more controlled nature of use and a focus on achieving therapeutic effects, and thus a reduced likelihood of engaging in high-risk use patterns [16].

In short, while a significant portion of the population – perhaps a third of the population – can be described as cannabis users, only a relatively small portion of cannabis users – less than 10% of the population (a quarter of past-year users), and perhaps substantially less when medical use is taken into account – should be considered at-risk for serious cannabis-induced medical problems.

Are researchers accurately identifying cannabis users in their studies?

When conducting a study on a particular adverse health effect associated with cannabis use, researchers generally start with a study population and then divide that population into a study group of cannabis users and a control group of non-users. The researchers then compare the prevalence of adverse health effects for the issue being studied in the study group to those in the control group, and if the prevalence of effects is greater in the study group, researchers conclude that cannabis use is associated with that adverse effect. Yet, in the vast majority of cases, many of the participants included in the control group (non-cannabis users) actually belong in the study group (cannabis users), where it is generally the less intensive users who are inappropriately classified as being non-users. As a result, the reported chances that average cannabis users may experience adverse health effects tend to be biased upward. In these cases, correcting the misidentification would reduce – if not entirely eliminate – the reported differences in the prevalence of effects between cannabis users and non-users.

A simple example will help elucidate how researchers' failure to accurately identify cannabis users among study participants leads to biased conclusions.

A simple example

Many studies tend to have large control groups relative to the size of the study group, that is, the number of non-cannabis users is generally large with respect to the number of identified cannabis users. So now make the following assumptions for the example, which are illustrated in Figure 3:
- There are 100 participants in our study example.
- Thirty of the 100 participants (30%) are actual past-year cannabis users (as per MTF).
- Only 8 of these 30 (27%) are identified as cannabis users and placed in the study group, leaving 92 participants in the control group.
- The remaining 22 actual cannabis users are mistakenly placed in the control group with the actual 70 non-cannabis users, for a total of 92 participants in the control group.
- Two of the 8 (25%) participants in the study group are found to have the adverse health effect being studied, 4/22 (18%) misidentified cannabis users in the control group have the adverse health effect, and 10/70 (14%) non-cannabis users in the control group have the adverse health effect.

When the study participants are correctly identified, with all the actual cannabis users placed in the study group and actual non-cannabis users placed in the control group, we get the depiction in Figure 4.

Table 1 provides a comparison of the outcomes under the assumptions used when cannabis users are misidentified with the outcomes when cannabis users are correctly identified.

The odds ratio is the increase in probability of experiencing the adverse effect for cannabis users relative to non-users. In the example, cannabis users are estimated to be 64% more likely to experience the adverse effect when cannabis users are misidentified as non-users, whereas they are estimated to be only 40% more likely to experience the adverse effect when cannabis users are correctly classified.

More generally, when
(i) the number of misidentified users is large, and/or
(ii) the probability that misidentified users experiencing the adverse effect is more similar to that for non-users,

then accurately classifying misidentified cannabis users will likely significantly reduce – if not altogether eliminate – any differences in the probability of experiencing adverse effects between users and non-users. Many studies that use diagnosis codes or self-reported information to identify cannabis and non-cannabis users involve large sample sizes. These studies are likely to have large numbers of misidentified cannabis users and, thus, invalid reported outcomes.

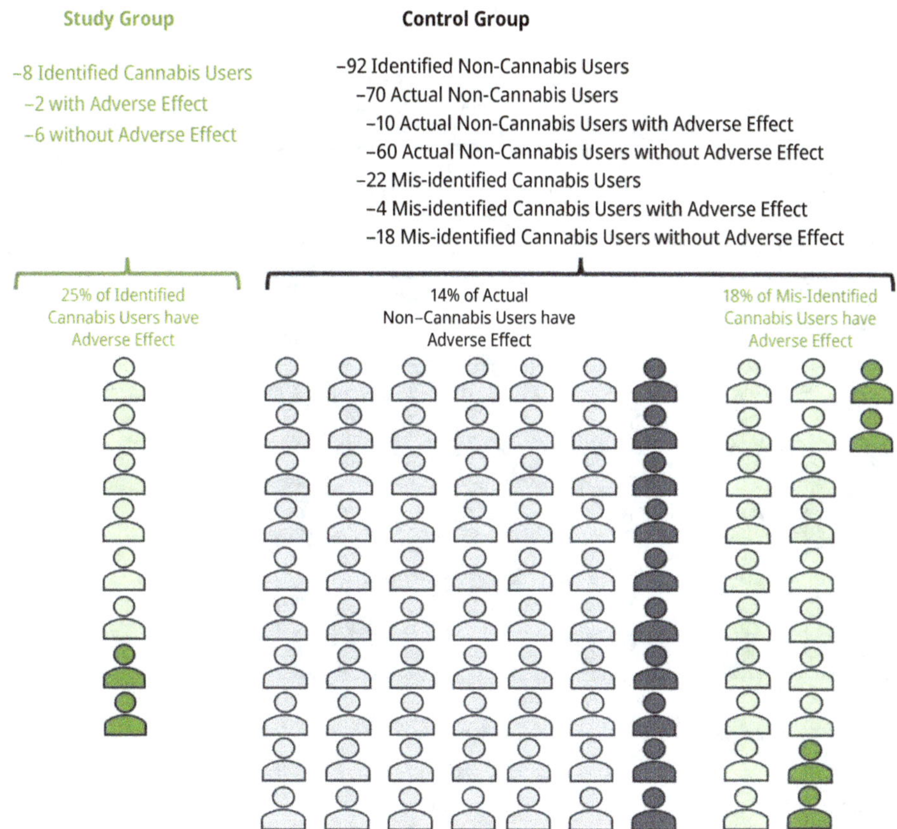

Figure 3: Example of a study with misidentified cannabis users.

At the same time, the misidentification of users could also cause reported outcomes to overestimate the probability of non-cannabis users experiencing adverse effects, if the misidentified cannabis users are closer to identified cannabis users than to non-users in the likelihood of experiencing adverse effects. Notice that in the example, the probability that non-users experience a negative effect decreases from 15% to 14% when misidentified users are correctly classified (Table 1).

The big question, then, is this: are the health effects of misidentified cannabis users more similar to those of non-users or to those of users? To answer this question, we need to know more about the misidentification problem. But first, let's see the problem in action.

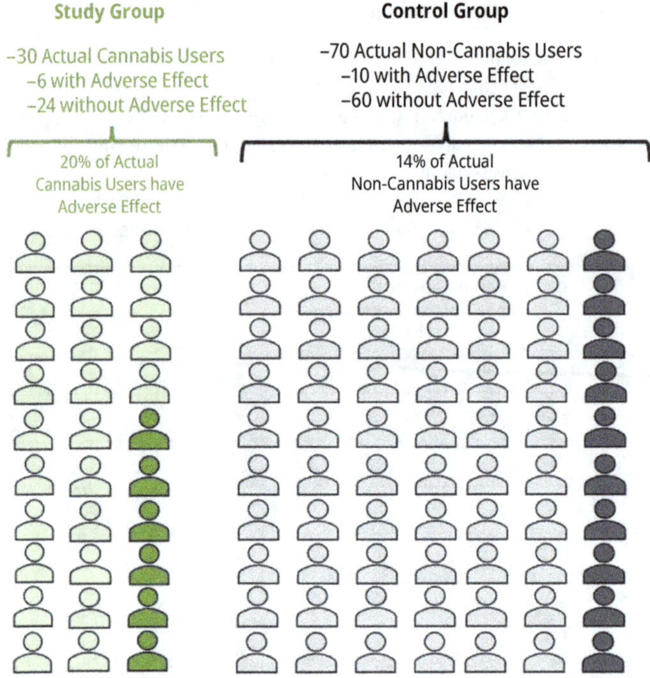

Figure 4: Example with accurately identified cannabis users.

Table 1: Odds ratios when cannabis users are misidentified versus accurately identified.

	Cannabis users are misidentified		Cannabis users are accurately identified	
	Number	Percent	Number	Percent
Total cannabis users	8		30	
With adverse outcome	2	25.0%	6	20.0%
Without adverse outcome	6	75.0%	24	80.0%
Total non-cannabis users	92		70	
With adverse outcome	14	15.2%	10	14.3%
Without adverse outcome	78	84.8%	60	85.7%
Odds ratio		1.64		1.40

Methods for identifying cannabis users

When partitioning the population into cannabis users and non-users, researchers generally employ one of three methods for identifying cannabis users: diagnosis codes

(ICD-9 or ICD-10 codes) contained in patients' medical or health insurance records, biological testing (e.g., urine, blood, and hair), or self-reported information from study participants. Yet, each of these three methods leads researchers to underidentify cannabis users. This section presents examples of how each of these methods fails to adequately capture all cannabis users in study populations.

Using diagnosis codes to distinguish cannabis users

Studies of the impact of using cannabis on health outcomes that analyze either patients' hospital records or their health insurance claims use diagnosis codes (ICD-9-CM or ICD-10-CM) to identify cannabis users. However, ICD-9 and ICD-10 codes only exist to diagnose cannabis use disorders, that is, there are no diagnosis codes for people who use, but do not abuse, cannabis.

As previously noted, the CDC estimates that 3 in 10 cannabis users have been diagnosed with CUD. Since there are no diagnosis codes for cannabis users without CUD, the majority of such users – the 70% or more cannabis users without CUD – will not be captured in the healthcare system. It follows that any study that relies solely on ICD-9-CM or ICD-10-CM codes for identification will fail to correctly identify the vast majority of cannabis users.

Take the example of a 2020 study by researchers at the CDC who compared the prevalence of mold infections in cannabis users to those in non-users by investigating health insurance claims using ICD-10-CM codes to identify cannabis users [17]:

> We identified patients with ICD-10-CM codes for mold infections (aspergillosis [B44], mucormycosis [B46]) and certain other fungal infections (blastomycosis [B40], coccidioidomycosis [B38], cryptococcosis [B45], histoplasmosis [B39]) among persons who used cannabis (F12.1, F12.2, F12.9) and persons who did not use cannabis. We further explored differences between ICD-10-CM codes for cannabis abuse or dependence (F12.1 and F12.2) and unspecified cannabis use (i.e., without mention of abuse or dependence) (F12.9). [17]

According to ICD10Data.com, F12 is the diagnosis code for "cannabis-related disorders," specifically, "excessive use of marijuana with associated psychological symptoms and impairment in social or occupational functioning" [18]. The researchers suggest that F12.9 captures cannabis users who do not abuse cannabis; however, the coding guidelines indicate F12.9 falls within the umbrella of F12, "cannabis-related disorders," which falls under the umbrella of F01–F99, "Mental, Behavioral and Neurodevelopmental disorders." The diagnosis codes employed by the researchers to identify cannabis users thus fail to capture all cannabis users who do not abuse cannabis.

Notably, researchers found health insurance claims for 53,217 patients with codes for cannabis disorders and 21,559,558 claims for patients without such codes, yielding an identified prevalence of cannabis use within this patient population of 0.25%. The researchers themselves noted that cannabis use may have been underreported in the health claims and that the US population reports about 9% cannabis use. By their very own admission, the study group captures less than 3% (=0.25%/9%) of actual cannabis users, but the researchers do not suggest this significant underidentification may have invalidated the results of the study.

The researchers report that 40 of the 53,217 identified cannabis patients (0.075%) were diagnosed with mold infections, while 6,294 of the 21,559,558 non-cannabis patients (0.029%) were diagnosed with mold infections. The researchers conclude that cannabis patients were 2.6 (=0.075%/0.029%) times as likely to suffer from a mold infection (Table 2). However, if you assume 9% of all the patients in their sample were actually cannabis users as they suggest – though this likely understates actual cannabis use in the population – and if you assume the misidentified cannabis users were just as likely to suffer from a mold infection as non-cannabis users, then the odds ratio drops from 2.6 down to 1. In other words, it is quite possible that if the researchers had accurately identified all cannabis users in the health claims, then they would have found no difference between rates of mold infections in cannabis users and those in non-cannabis users.

Table 2: Fungal infections in cannabis patients identified in health insurance claims.

	Identified cannabis users	Unidentified cannabis users	Total cannabis users	Not cannabis users	Total	Odds ratio
Number of patients	53,217		53,217	21,559,558	21,612,775	
Number of patients with fungal infections	40		40	6,294	6,334	
Percent of patients with fungal infections	0.075%		0.075%	0.029%		2.6
Adjusted number of patients	53,217	1,891,933	1,945,150	19,667,625	21,612,775	
Adjusted number of patients with fungal infections	40	552	592	5,742	6,334	
Adjusted percent of patients with fungal infections	0.075%	0.029%	0.030%	0.029%		1.0

Source: Benedict and Thompson [17].

Using self-reporting and biological testing to distinguish cannabis users

Some studies ask participants about their cannabis use, while also using some form of biological testing (e.g., urine, blood, and hair) to determine whether or not those studies' participants have cannabis metabolites in their systems. Researchers then compare the self-reports of cannabis use with the biological tests for cannabis metabolites to assess the accuracy of both modes of collecting cannabis use information. These studies generally reveal large discrepancies between participants who self-report using cannabis and those who test positive for cannabis use. These discrepancies suggest that both self-reported information and biological testing significantly underidentify cannabis users.

Consider two studies of the prevalence of cannabis use in pregnant women that compared self-reported information with outcomes generated from biological testing. While the number of participants in the two studies differed by several orders of magnitude, and one study used blood tests while the other used urine tests, the outcomes from these two studies were remarkably similar.

In the first study, Shiono et al. [19] examined the impact of cannabis and cocaine use on birth outcomes. The study involved women from seven different prenatal clinics during the 1984–1989 period. In addition to other demographic and lifestyle factors, participants were asked:

> At any time during this pregnancy have you used any of the following? Drugs to control nausea or vomiting, tranquilizers or sedatives, amphetamines, marijuana, methadone, heroin, LSD, PCP, cocaine, other. [19]

Nurses also obtained blood serum from study participants, which were tested for cannabis metabolites.

For the second study, Young-Wolff et al. [20],

> The primary study sample comprised KPNC [Kaiser Permanente Northern California] pregnant women who were screened for self-reported cannabis use during pregnancy [as part of standard prenatal care] between 2009 and 2017, and had a urine toxicology test for cannabis 2 weeks from the date they completed the self-reported screening questionnaire. [20]

In both studies the vast majority of the participants reported no cannabis use and also tested negative for cannabis use, 89% in Shiono et al. and 94% in Young-Wolff et al. (Table 3). About 6% of the participants in Shiono et al. reported using cannabis, compared with 3% in Young-Wolff et al., while 8% versus 5%, respectively, tested positive for cannabis (Table 3).

In Shiono et al., 43% of the participants who reported using cannabis also tested positive, compared with 66% in the Young-Wolff study (Figure 5). These are measures of the positive predictive value of the biological tests, showing the biological tests correctly identified less than half of reported users in Shiono et al. and two-thirds of reported users in Young-Wolff et al.

Interestingly, Young-Wolff et al. noted that "Clinicians in California are not required to alert child protective services or law enforcement agencies when a woman uses substances during pregnancy," so patients can be assured they will not be penalized for admitting to using cannabis during their pregnancy. No such assurances were given to women in Shiono et al., so we might expect women in Young-Wolff et al. to be more forthcoming about admitting to cannabis use than women in Shiono et al. Yet, the two studies are nearly identical in the portion of women testing positive for cannabis metabolites who also report using cannabis (31% vs. 34%) (Figure 6). These are measures of the sensitivity of reported use, where higher sensitivity suggests fewer false negatives, that is, fewer missed users. The relatively low sensitivities suggest that large portions of users (two-thirds or more in both studies) are missed with self-reported use.

Granted, pregnant women have greater incentives than women who are not pregnant to understate their cannabis use. That is, these studies may exaggerate the extent of underreporting of cannabis use we would find in the general population. Unfortunately, most studies comparing self-reported cannabis use to biological tests involve individuals in atypical situations, such as those in drug treatment programs [21] or those involved in motor vehicle accidents [22]. That is, in most studies comparing self-reported cannabis use to biological testing for cannabis use, participants have greater-than-average incentives to understate cannabis use. Nonetheless, the striking similarities in outcomes across the two studies, together with the potential magnitude of the problem, are suggestive.

Table 3: Cannabis use during pregnancy: reported use versus biological tests.

		Blood serum tests for cannabis			Urine tests for cannabis		
Number of participants		Tested +	Tested –	Total	Tested +	Tested –	Total
Reported cannabis use?	Yes	180 (2%)	237 (3%)	417 (6%)	4,707 (2%)	2,442 (1%)	7,149 (3%)
	No	405 (5%)	6,648 (89%)	7,053 (94%)	9,212 (3%)	264,664 (94%)	273,876 (97%)
	Total	585 (8%)	6,885 (92%)	7,470 (100%)	13,919 (5%)	267,106 (95%)	281,025 (100%)
Sources:		Shiono et al. [19].			Young-Wolff et al. [20].		

In short, studies strongly suggest that using diagnosis codes, self-reported information, and biological testing all significantly underidentify cannabis users in study populations.

Are misidentified users chronic users or causal users?

Indeed, research suggests that health outcomes of casual (e.g., monthly) cannabis users are more similar to those of non-users than they are to those of chronic (e.g., daily) users, since infrequent use minimizes risks of dependency [23], mental health disorders [24], cardiovascular disease [2], and respiratory harm [25]. So, what is left is to determine if the misidentified cannabis users identified in cannabis studies are more likely to be chronic users or casual users. If it is the latter, then we can conclude

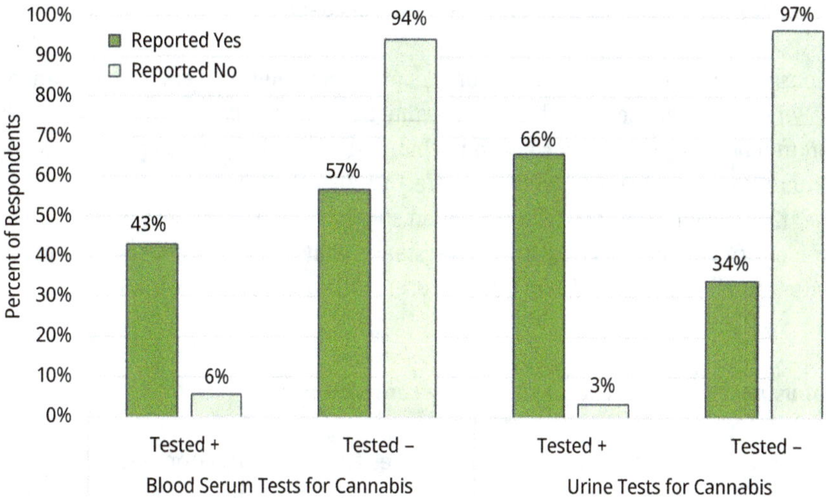

Figure 5: Cannabis use during pregnancy – Distribution of participants as a percent of self-reports. Sources: Shiono et al. [19] and Young-Wolff et al. [20].

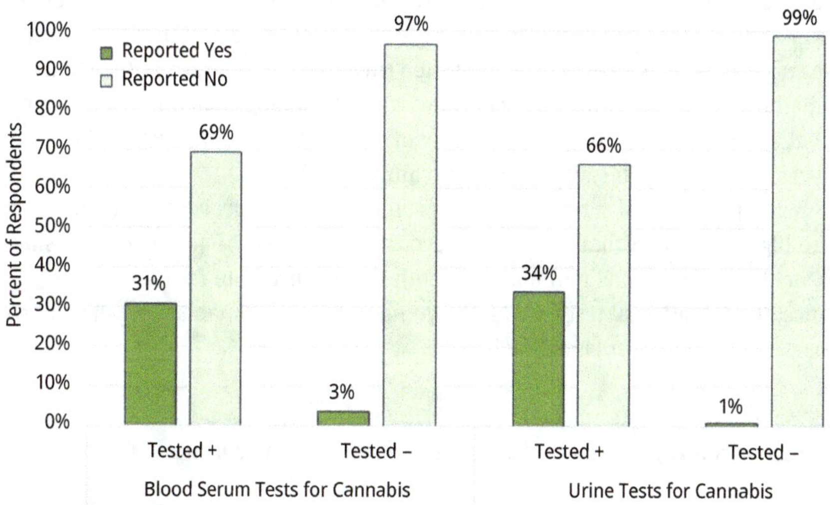

Figure 6: Cannabis use during pregnancy – distribution of participants as a percent of testing outcomes. Sources: Shiono et al. [19] and Young-Wolff et al. [20].

that studies on the adverse effects of cannabis use are likely only relevant for chronic users, while overstating, perhaps significantly, the risks to more casual cannabis users of adverse health effects.

Biases of using diagnosis codes to identify cannabis users

As previously discussed, diagnosis codes only capture cannabis abusers, but not cannabis users who do not abuse cannabis. At the same time, cannabis users diagnosed with CUD tend to have a high prevalence of comorbidities: "Cannabis use disorder was associated with other substance use disorders, affective disorders, anxiety, and personality disorders" [23]. Taken together, the information suggests that studies that use diagnosis codes to identify cannabis users tend to overstate the likelihood of adverse outcomes not only for casual users, but also for chronic users who do not abuse cannabis.

Biases of using biological testing to identify cannabis users

Biological testing may yield inaccurate outcomes in either direction, false negative (failure to capture) or false positive (false detection) for cannabis use, depending on a combination of individual-specific and test-specific factors. Detectability factors relating to the individual and substance being assessed include the type of drug being assessed, the size of the dose, the frequency of use, the route of administration, and differences in individual drug metabolism. Detectability factors relating to the testing process include the time the sample was collected relative to when the individual used the substance, the biological sample tested (urine, blood, hair, etc.), the choice of testing method used (immunoassay screening, gas chromatography/mass spectrometry, etc.), the metabolites being, the sensitivity of the method to detection, and the cutoff levels employed for determining positive cannabis use [21].

The high specificity of modern testing protocols tend to prevent false positives, while the high likelihood that testing will occur after enough time has passed since last use, or that last use was minimal, will both tend to promote false negatives [26]. On balance, then, biological testing will likely undercapture cannabis use, particularly casual use, rather than overcapturing it.

Biases of using self-reported information to identify cannabis users

Many studies have been conducted to help characterize the nature of self-reported cannabis use. Specific findings indicate that the accuracy of self-reported cannabis use tends to be affected by the following.
- **Stigma**: Greater social stigmas among the local population associated with using cannabis will decrease the accuracy of self-reported use [21, 27, 28].
- **Fear of punishment**: Greater fear of punitive consequences (e.g., legal sanctions), especially in high-risk populations, such as pregnant women, will decrease the accuracy of self-reported use [21, 27, 28]. Notably, many states have laws that penalize drug use by pregnant women [20, 28].

- **Setting**: Self-report was found to differ depending on the setting in which the individual is interviewed. For example, self-reports by individuals in drug treatment centers have been found to be more reliable upon entry than at 6-month or 9-month follow-up [28].
- **Interviewer**: Accuracy of self-report was found to differ depending on who administers the survey [27]. Self-report was found to be more reliable when conducted by researchers rather than by healthcare providers or by self-administered surveys [29].
- **Population**: Accuracy of self-report was found to differ by race [27] and be more reliable in individuals with a known history of drug use and/or enrolled in drug treatment programs [29].
- **Frequency of use**: Self-report was found to be less reliable with more frequent drug use [27, 28].
- **Date of use**: People significantly underreport the date of last use [28]. More generally, the telescoping effect is a commonly recognized cognitive bias that causes people to misremember the timing of events [27, 30].

As a whole, the evidence suggests that self-reports tend to undercapture cannabis use, especially less frequent use.

Conclusion

So much bad cannabis research is published because researchers cannot meaningfully and accurately distinguish cannabis users from non-users. The first part of the problem is the lack of a clear understanding among researchers and healthcare professionals as to what constitutes clinically meaningful cannabis use. The more serious problems commonly associated with cannabis use can be said to involve daily or near-daily use. While a relatively large portion of the population, perhaps a third, can be described as cannabis users, at most a quarter of those are daily or near-daily users. So it is likely that only a relatively small portion of cannabis users are at risk for more serious adverse effects. At the same time, the question is still wide open for debate. Cannabis products are heterogeneous in nature, they have different effects on different individuals, depending not only on lifetime use patterns, but also on the individual make up of cannabis users, and users tend to be a heterogeneous group. Before good cannabis research can be published, researchers must first come to agreement on what constitutes clinically meaningful cannabis use.

The second part of the problem is the inability of researchers to accurately identify cannabis users in their study populations. The various methods used to capture cannabis users – healthcare diagnosis codes, biological testing, and self-identification – have all been shown to significantly underidentify cannabis users as a whole, while overrepresenting higher intensity cannabis users and abusers. Consequently, reported outcomes tend to overstate risks not only to more casual cannabis users, but also to more intense users who do not abuse cannabis.

Before good cannabis research can be published, then, researchers must figure out how to adequately identify appropriately defined cannabis users. Having study participants self-identify is the most common method of identification. Yet, unless or until the stigma surrounding cannabis use is removed, and/or researchers can guarantee participants a safe reporting environment, participants will continue to underreport cannabis use, causing significant inaccuracies in reported outcomes.

References

[1]   National Institute on Drug Abuse. (2018, June). Marijuana. https://nida.nih.gov/sites/default/files/1380-marijuana.pdf

[2]   Jeffers, A.M., Glantz, S., Byers, A.L., Keyhani, S. (2024). Association of cannabis use with cardiovascular outcomes among US adults. Journal of the American Heart Association, 13(5), e030178.

[3]   Frysh, P., Booth, S. (2024, December 23). How marijuana affects your mind & body. WebMD. https://www.webmd.com/mental-health/addiction/marijuana-use-and-its-effects

[4]   National Academies of Sciences, Engineering, and Medicine; Health and Medicine Division; Board on Population Health and Public Health Practice; Committee on the Health Effects of Marijuana. (2017). The Health Effects of Cannabis and Cannabinoids: The Current State of Evidence and Recommendations for Research. National Academies Press.

[5]   Meier, M.H., Caspi, A., Ambler, A., Harrington, H., Houts, R., Keefe, R.S., McDonald, K., Ward, A., Poulton, R., Moffitt, T.E. (2012). Persistent cannabis users show neuropsychological decline from childhood to midlife. Proceedings of the National Academy of Sciences, 109(40), E2657–E2664.

[6]   Kaplan, A.G. (2021). Cannabis and lung health: Does the bad outweigh the good?. Pulmonary Therapy, 7(2), 395–408.

[7]   Hill, K.P. (2015). Medical marijuana for treatment of chronic pain and other medical and psychiatric problems: A clinical review. JAMA, 313(24), 2474–2483.

[8]   Volkow, N.D., Swanson, J.M., Evins, A.E., DeLisi, L.E., Meier, M.H., Gonzalez, R., Bloomfield, M.A., Curran, H.V., Baler, R. (2016). Effects of cannabis use on human behavior, including cognition, motivation, and psychosis: A review. JAMA Psychiatry, 73(3), 292–297.

[9]   Caspi, A., Moffitt, T.E., Cannon, M., McClay, J., Murray, R., Harrington, H., Taylor, A., Arseneault, L., Williams, B., Braithwaite, A., Poulton, R., Craig, I.W. (2005). Moderation of the effect of adolescent-onset cannabis use on adult psychosis by a functional polymorphism in the catechol-O-methyltransferase gene. Biological Psychiatry, 57(10), 1117–1127.

[10] Wainberg, M., Jacobs, G.R., Di Forti, M., Tripathy, S.J. (2021). Cannabis, schizophrenia genetic risk, and psychotic experiences: A cross-sectional study of 109,308 participants from the UK Biobank. Translational Psychiatry, 11(1), 211.

[11] Patrick, M.E., Miech, R.A., Johnston, L.D., O'Malley, P.M. (2024). Monitoring the Future Panel Study Annual Report: National Data on Substance Use among Adults Ages 19 to 65. 1976–2023, Institute for Social Research, University of Michigan.

[12] Patel, J., Marwaha, R. (2025). Cannabis Use Disorder. In: StatPearls. StatPearls Publishing, https://www.ncbi.nlm.nih.gov/books/NBK538131/.

[13] Centers for Disease Control and Prevention. (2025, March 7). Marijuana and public health: Data and statistics. https://www.cdc.gov/cannabis/data-research/facts-stats/

[14]   Fisher, R. (2024). Do 30% of cannabis users really have cannabis use disorder? Quantaa. https://quantaa.com/blog/597-do-30-of-cannabis-users-really-have-cannabis-use-disorder

[15]   Compton, W.M., Dawson, D.A., Goldstein, R.B., Grant, B.F. (2013). Crosswalk between DSM-IV dependence and DSM-5 substance use disorders for opioids, cannabis, cocaine, and alcohol. Drug and Alcohol Dependence, 132(1–2), 387–390.

[16]   Dai, H., Richter, K.P. (2019). A national survey of marijuana use among US adults with medical conditions, 2016–2017. JAMA Network Open, 2(9), e1911936.

[17]   Benedict, K., Thompson, G.R. III, Jackson, B.R. (2020). Cannabis use and fungal infections in a commercially insured population, United States, 2016. Emerging Infectious Diseases, 26(6), 1308–1310.

[18]   ICD10Data.com. (2024). ICD-10-CM diagnosis code F12. https://www.icd10data.com/ICD10CM/Codes/F01-F99/F10-F19/F12-/F12

[19]   Shiono, P.H., Klebanoff, M.A., Nugent, R.P., Cotch, M.F., Wilkins, D.G., Rollins, D.E., Carey, J.C., Behrman, R.E. (1995). The impact of cocaine and marijuana use on low birth weight and preterm birth: A multicenter study. American Journal of Obstetrics and Gynecology, 172(1), 19–27.

[20]   Young-Wolff, K.C., Sarovar, V., Tucker, L.Y., Goler, N., Conway, A., Weisner, C., Armstrong, M.A., Alexeeff, S. (2020). Validity of self-reported cannabis use among pregnant females in northern California. Journal of Addiction Medicine, 14(4), 287–292.

[21]   Buchan, B.J., Dennis, M.L., Tims, F.M., Diamond, G.S. (2002). Cannabis use: Consistency and validity of self-report, on-site urine testing, and laboratory testing. Addiction, 97(Suppl. 1), 98–108.

[22]   Salottolo, K., McGuire, E., Madayag, R., Tanner, A.H. II, Carrick, M.M., Bar-Or, D. (2022). Validity between self-report and biochemical testing of cannabis and drugs among patients with traumatic injury: Brief report. Journal of Cannabis Research, 4(1), 29.

[23]   Hasin, D.S., Kerridge, B.T., Saha, T.D., Huang, B., Pickering, R., Smith, S.M., Jung, J., Zhang, H., Grant, B.F. (2016). Prevalence and correlates of DSM-5 cannabis use disorder, 2012–2013: Findings from the National Epidemiologic Survey on Alcohol and Related Conditions–III. American Journal of Psychiatry, 173(6), 588–599.

[24]   Lev-Ran, S., Roerecke, M., Le Foll, B., George, T.P., McKenzie, K., Rehm, J. (2014). The association between cannabis use and depression: A systematic review and meta-analysis of longitudinal studies. Psychological Medicine, 44(4), 797–810.

[25]   Tashkin, D.P. (2013). Effects of marijuana smoking on the lung. Annals of the American Thoracic Society, 10(3), 239–247.

[26]   Goodwin, R.S., Darwin, W.D., Chiang, C.N., Shih, M., Li, S.H., Huestis, M.A. (2008). Urinary elimination of 11-nor-9-carboxy-delta9-tetrahydrocannabinol in cannabis users during continuously monitored abstinence. Journal of Analytical Toxicology, 32(8), 562–569.

[27]   Johnson, T., Fendrich, M. (2005). Modeling sources of self-report bias in a survey of drug use epidemiology. Annals of Epidemiology, 15(5), 381–389.

[28]   Garg, M., Garrison, L., Leeman, L., Hamidovic, A., Borrego, M., Rayburn, W.F., Bakhireva, L. (2016). Validity of self-reported drug use information among pregnant women. Maternal and Child Health Journal, 20(1), 41–47.

[29]   Skelton, K.R., Donahue, E., Benjamin-Neelon, S.E. (2022). Validity of self-report measures of cannabis use compared to biological samples among women of reproductive age: A scoping review. BMC Pregnancy and Childbirth, 22(344).

[30]   Janssen, S.M., Chessa, A.G., Murre, J.M. (2006). Memory for time: How people date events. Memory & Cognition, 34(1), 138–147.

Andrew M. Peterson and Jahan Marcu

Pandora's plant: a series of entirely predictable absurdities

Introduction: the cannabis paradox landscape

The debates over cannabis can lull us into a glaze of policy disputes coated onto modern myths, echoing the tale of Pandora and the Greek gods. Pandora's box was kept closed – sealed for years by prohibition, regulation, and stigma. Now, with state-level legalization, that box has been opened, releasing both potential harm and hope to millions. Unlike the Greek fable, however, our modern version lacks a clear, guiding narrative; we do not fully understand what has been unleashed – or why we unleashed it in the way that we did. Research trails far behind public use, and perception outpaces knowledge.

A harsh truth emerges from the collective contradictions: societal acceptance of cannabis has made significant strides, yet multiple frontiers of resistance remain (see Box 8.1). Workplaces, healthcare systems, regulatory frameworks, and research establishments cling to outdated policies, treating cannabis use as evidence of incompetence or moral failing, regardless of context or legality. The plant is lawful in one hand under certain circumstances, criminal in another under different circumstances.

Box 8.1: By the numbers: the scope of cannabis in America

Usage: About 52.5 million Americans (19%) used cannabis at least once in 2021, with 15% reporting active use in 2023–2024. More than 1 in 3 women over 21 consume cannabis, and daily or near-daily cannabis users (17.7 million) now outnumber daily alcohol users (14.7 million) for the first time [1, 2].

Legal status: As of June 2025, 48 states allow some form of medical cannabis use, while 24 states allow adult recreational use [3]. Seventy-four percent of Americans live in a state where cannabis is legal for either recreational or medical use, and 79% live in a county with at least one cannabis dispensary [4].

Economic impact: The US cannabis industry is expected to reach $45 billion in 2025, adding $115.2 billion to the economy in 2024, while supporting 440,445 full-time jobs [4].

Criminal justice: Despite widespread legalization, over 200,000 Americans were arrested for cannabis violations in 2023 [5], with federal agents making 5,764 marijuana-related arrests in 2024 [6]. Black Americans are arrested for cannabis possession at nearly 4 times the rate of white Americans [6].

Workplace impact: 4.3% of general workers tested positive for cannabis in 2022 – the highest level ever recorded, while at least nine legalization states have employment protections, though most workers remain vulnerable to termination.

One example of the prevailing contradictions is drug testing – the use is lawful but a positive result in blood or urine can result in punishments of all types depending on the circumstances such as driving, being a parent, or workplace policies. As more states reform their cannabis laws, it's imperative for society to evolve its perceptions and align its policies with current realities. Until then, the various stigmas surrounding *Cannabis* will continue to force individuals into difficult choices between their health and employment careers, between dignity and compliance.

This chapter will examine what might be called the *Cannabis Paradox Landscape*: a terrain where acceptance coexists with stigma, progress is shadowed by ignorance, and science is outpaced by politics. We will navigate seven of its most glaring paradoxes: the Knowledge Paradox, the Safety Information Paradox, the Legal Paradox, the Decriminalization Paradox, the Equity Paradox, the Future Paradox, and the Political Paradox.

Within this framework, we will explore the entanglement of law, research and funding, equity, marketing, public health and politics. Each paradox, in its own way, illuminates not just what cannabis is, but what society insists on making it.

Knowledge paradox

The knowledge paradox
Core contradiction: Millions use cannabis medicinally despite lack of traditional research data, while federal restrictions prevent the very studies needed to validate or refute these uses.

When something appears to be a "paradox," it generally indicates a choice between two paths for resolution:
1. undertake the difficult work of reconciliation, or
2. abandon the myths and assumptions that led to the paradox and search for a new paradigm or predictive model.

This chapter uses a unique, mixed-discipline approach (MDA) of oral history and research to outline the most pressing paradoxes and contradictions regarding cannabis use in the United States:

> "This is the paradox. Regardless of what the science tells us, people will continue to consume cannabis." – Roger Ladouceur, MD MSc, CCMF(SP), FCMF, 2018 [7]

The intersection of policy, medical practice, and scientific research presents one of the most contentious challenges in modern healthcare. As cannabis transitions from a federally prohibited substance to state-sanctioned medicine, clinicians and patients navigate uncharted territory where the rules of evidence-based medicine collide with the immediacy of patient demand:

> "Medical research has allowed statistics to become the supreme judge of its inventions . . . We will put it bluntly: Are American doctors going to let people die to satisfy the bureau of drugs' chi-square studies?" – Daniel Carpenter, Reputation and Power [8]

For decades, the debate over the medical benefits has persisted with little resolution. The lack of ability to research the product due to federal regulations ensures definitive answers remain out of reach. This is just one systemic barrier blocking researchers from answering efficacy questions. Despite good data showing cannabis has efficacy in limited disease states and poses potential harms, people still use it for many different conditions. The disconnect is glaring: the science inches forward, while human behavior runs ahead. Public health messaging and policy, caught in the middle, are left to manage a population that refuses to wait for consensus.

Federal funding remains heavily restricted and subject to abrupt policy-driven disruptions. Researchers must navigate unpredictable grant freezes and reductions, forcing many to abandon studies altogether. The barriers are not merely financial. Regulatory labyrinths stretch timelines, inflate costs, and discourage institutions from participating at all. The ongoing debate about what counts as valid cannabis science is exacerbated by regulatory complexities:

> "Without fully funded indirect costs, 'I literally cannot do my research' . . . 'We're all terrified that our work will grind to a halt.'" – Angela Bryan, University of Colorado Boulder as cited in Roberts [9]

> "Every step [in cannabis clinical research] involved protracted timelines, unexpected costs, and multiple layers of institutional and federal review . . . These experiences illustrate why so few academic centers are willing or able to pursue cannabis-related clinical research despite widespread clinical interest." – Dr. Morenike Kheirbek as cited in Marijuana Moment [10]

> "Statutory restrictions on what can be studied and a mandate to oppose any attempts to reschedule substances like cannabis make no sense. It's time to update the law to reflect the current use of cannabis in the United States and its medical benefits. The federal government needs to catch up to the states." – Congresswoman Dina Titus, (D-Nevada), 2025 [11]

This is the knowledge paradox in its starkest form: patients, physicians, and scientists are left in a fog of uncertainty, while the law demands certainty as a precondition for progress. The plant moves faster than the policies; the people move faster than the research. The box has been opened, yet the tools to measure what has escaped remain locked away.

Safety paradox

The Safety information paradox
Core contradiction: Conflicting safety narratives from authorities create an information vacuum filled by biased sources, while the same data gets weaponized by both pro- and anti-cannabis advocates.

> "I often hear complaints from other doctors that there isn't adequate evidence to recommend medical marijuana, but there is even less scientific evidence against it, because not enough research has been done." – Peter Grinspoon, 2020 [12]

Research exists, but much of it has been pointed down a narrow corridor – money poured into cataloging harms, sometimes so aggressively it distorts the science – while real questions of risk-benefit ratios leave patients to rely on anecdote, advertising, and advocacy slogans. Federal funding remains heavily restricted, and what research does exist seems so often misinterpreted or stripped of nuance by media outlets hungry for headlines.

> "Marijuana is not a benign drug. As Surgeon General, I urge other physicians and professionals to advise parents and patients about the harmful effects of using marijuana and to urge discontinuation of its use"
>
> – C. Everett Koop, "Surgeon General's Advisory on Marijuana," 1982 [13]

From C. Everett Koop's recommendation in 1982 to Joycelyn Elder's 2004 statement cited in Adali's publication that:

> "Indeed, marijuana is less toxic than many of the drugs that physicians prescribe every day" [14]

the US Surgeon General's office has always had mixed feelings about marijuana use. These conflicting narratives reflect broader tensions between science, politics, and public opinion.

Compounding this polarized viewpoint is product safety. Investigations reveal that both legal and illicit products can harbor mold, pesticides, and heavy metals. While dispensary products tend to be safer, the legal supply is not immune to contamination and adulteration. In 2023, CNN headlined that "Marijuana Users have more heavy metals in their bodies," based on a national survey that didn't distinguish between regulated and unregulated sources [15]. The result: a headline that could be wielded by prohibitionists as proof of danger or dismissed by advocates as irrelevant. The data itself became less a finding than a weapon, shaped to fit whomever wields it.

This exemplifies misinformation in the *Cannabis* world. The oft lack of distinction between regulated and unregulated sources creates opportunities for disinformation. One side can claim all marijuana is inherently contaminated while the other dismisses findings as irrelevant to legal products. These data become tools for both camps to selectively interpret evidence supporting predetermined positions.

Patients seeking medical cannabis frequently voice frustration about the lack of reliable, unbiased information regarding product safety, dosing, and efficacy. While many turn to cannabis as an alternative to conventional pain medications, often perceiving it as safer than opioids or antidepressants, users remain acutely frustrated by the lack of reliable, unbiased guidance on safety, dosing, and efficacy. A 2023 study by Kleidon et al. sheds light on how older adults – one of the fastest-growing groups of medical cannabis users – navigate safety and information gaps [16]. The survey of adults 65 and older in Philadelphia found 76% considered cannabis "a highly important treatment" for older adults, yet most relied on the internet or social media rather than healthcare providers for information, with just 23% reporting their primary care provider asked about *Cannabis* use in the past year. Even among the medically vulnerable, the conversation is outsourced to a search engine or generative AI, not the doctor's office. This disconnect illustrates how confusion about the risks and benefits of cannabis is compounded by a lack of authoritative, accessible guidance from medical professionals:

> "This pilot study highlights the need for accurate and reliable information about cannabis for older adults and their healthcare providers . . ." – Kleidon et al. [16]

The consumer perspective for recreational users is equally shaped by uncertainty. While many prefer the legal market for perceived safety and quality, surveys show the illicit market still thrives. Perceptions of variety, cost, and less barriers to access seem to outweigh assurances of safety. In a recent survey of more than 3,000 Swiss cannabis consumers, nearly two-thirds still reported obtaining cannabis from the illicit market [17].

The study concluded that "A carefully designed [regulated market] that takes into account the consumers' perspective is likely to transfer them to the regulated market and to engage vulnerable populations" [17].

Yet, regulatory patchwork and inconsistent state-level product testing standards in the United States mean that even compliant consumers may face exposure to unrecognized contaminants. Another 2025 report summarized the following.

> "Finding cannabis products that are safely sourced and properly regulated can be tricky . . . traces of other harmful substances might be lurking in a tasty-looking edible or nicely packaged vape." – NPR, 2025 [18]

In a haze of uncertainty, marketing, advocacy, and misinformation thrive – while patients and the public are left to choose between risk without knowledge or knowledge without trust:

> "As the cannabis industry continues to expand and attract a broader consumer base, addressing these regulatory inconsistencies will be essential to protecting public health and ensuring consumer safety." – Emma Wozniak, 2025 [19]

The legal paradox: the workplace trap

The legal paradox
Core contradiction: Cannabis can be legal to purchase and use, yet result in job loss, while social equity programs promise justice but often fail to deliver meaningful change to affected communities.

Nowhere is the contemporary cannabis paradox more brutal than in American workplaces: individuals may legally purchase adult-use products, use federally legal CBD products, or take prescription THC under medical supervision, yet still find themselves treated as intoxicated contraband traffickers. Employers – invoking the Drug-Free Workplace Act, safety standards, or zero-tolerance policies – retain broad discretion to terminate employees who use cannabis even outside work hours or for medical reasons.

Organizations with Drug-Free Workplace Act policies may terminate individuals using state-legal medical marijuana since it remains federally illegal. CBD's quasi-legal status provides no protection. Despite being marketed as non-intoxicating and federally lawful, CBD products can trigger positive drug tests, often due to the presence of trace amounts of THC. Prescription THC products like dronabinol offer no clearer path to security.

A real-world example: truck drivers under the Department of Transportation policy cannot use CBD as a defense for failing THC screens, even without impairment. This highlights a sub-paradox – organizations use measurements of past exposure (urine metabolites) to conclude workplace impairment. You may never have been impaired at work, and the THC might be legal, but you could be treated as an illicit drug user.

Here the paradox tightens its grip: legalization expands personal freedoms, but employment policies collapse them back into prohibition. Workers must navigate a landscape where what is lawful in public and private spaces can remain punishable in the workplace.

In the American workplace, cannabis is both legal and fireable – Schrodinger's drug – it's not misconduct until the urine is tested.

The decriminalization paradox: when definitions outrun devices

The decriminalization paradox
Core contradiction: Decriminalization promises simplicity and freedom through definition, yet in practice, it produces complexity and constraint through measurement. The very act of drawing legal boundaries (0.3% THC threshold) increased enforcement burdens because the boundary required tools, expertise, and precision that the law never provided.

It began, as these things often do, with optimism – the legislative kind, heavy with faith in definitions and light on calibration.

In 2018, Congress drew a line through a molecule and declared victory. Cannabis with less than 0.3% delta-9-THC would henceforth be "hemp," and everything above it would remain contraband. The idea was elegant, almost Newtonian: a single numeric threshold that would separate the lawful from the illicit, the farmer from the felon.

By all appearances, the system had finally found order. But numbers, like gods, are fickle when worshipped without instruments.

But science and bureaucracy are notoriously bad dance partners. A 2023 analysis in the *Journal of Criminal Justice* found that following the Farm Bill, police actually became less efficient. Cannabis incidents didn't disappear; they only took longer to resolve. Officers in Dallas, for instance, spent an additional thirty minutes per case after 2018, a figure that crept higher with each passing month. Fewer arrests, more time – a curious form of liberation, where freedom from cannabis meant still more paperwork about it. The law, meant to streamline justice, had jammed its own gears.

The problem was not malice but measurement. The Farm Bill's distinction was biochemical – 0.3% delta-9-THC by dry weight – a number so delicate it could evaporate in a humid room. Numbers, however, do not shimmer, smell, or smoke. You cannot see 0.3%. You cannot smell the difference between hemp and marijuana, whether it is blooming in a field or burning in a joint. Yet officers were asked to enforce that distinction with their eyes and instincts. The more precise the statute became, the less enforceable it was.

Without validated testing, proper calibration, and certified reference materials – each traffic stop became a hypothesis written on the back of a citation. The United States had decriminalized cannabis in principle, but criminalized uncertainty in practice.

Compounding the confusion was what the law forgot to count. In its zeal to define hemp, Congress focused on delta-9-THC – the molecule that headlines arrests – while ignoring its quiet precursor, THCA. In living plants, THCA is plentiful while delta-9-THC is scarce; a little heat, time, or sunlight turns one into the other. By ignoring this transformation, legislators effectively legalized potential energy. Farmers quickly realized they could cultivate crops rich in THCA – chemically primed to become intoxicating THC – yet still test "legal." Laboratories, many underfunded or inconsistent, struggled to track this conversion, and a new industry flourished in the liminal space between chemistry and statute. Once again, definition sprinted ahead of device; policy tried to fix nature in place.

By 2025, the paradox had ripened into crisis. The National Association of Attorneys General issued a rare bipartisan letter urging Congress to close the "hemp loophole," warning that intoxicating derivatives such as delta-8, delta-10, and THC-O had exploited the Farm Bill's wording to flood the market with psychoactive products that were technically lawful [20]. As of writing this, federal policy changes are looming that would ban all hemp products that have more 0.4 mg total THC per container, among other restrictions. If the Dallas study had captured the procedural confusion of 2018, the attorneys general captured its metastasis. What began as definitional drift had become regulatory paralysis. It remains unclear whether these alarms reflect genuine public-health concerns or simply the fiscal exhaustion of local law enforcement, state crime laboratories, and prosecutors seeking relief from an unfunded mandate – and perhaps the freedom to redirect scarce resources toward more immediate crises such as violent crime, sexual assault, and the fentanyl epidemic.

There is a peculiar rhythm to American drug policy. We legalize by definition and then spend a decade litigating what the definition means. The 0.3% threshold, borrowed from the 1970s Canadian hemp project never intended for US criminal code, was elevated from an agronomic footnote to federal scripture. And yet, that single decimal continues to dictate arrests, prosecutions, and entire markets. Slowly it dawned on lawmakers that THC is not a single molecule but a family – a spectrum of isomers and acids – and that the neat boundary they had drawn was an illusion. We built a legal framework that depends on chromatography but demands certainty from gut instinct. It is as if Congress legislated gravity and left enforcement to men with kites.

For the courts, the lesson is empirical humility: one cannot adjudicate molecules without measurement. For laboratories, the task is civic as much as technical: to become the infrastructure that policy forgot. And for those who traverse both worlds – science and law – the paradox is almost moral. Every reform carries within it the seed of its next contradiction. Decriminalization did not fail; it merely revealed that freedom, like precision, takes time.

The data tell a modest but eloquent story: when law moves faster than science, enforcement slows to a crawl. The 2018 Farm Bill promised clarity and delivered ambiguity; the 2025 attorneys general letter now asks Congress to rescue the system from its own definition. Between those two documents lies the central moral of this paradox: that progress, in policy as in chemistry, depends not on purity but on precision. Until instruments catch up with legislation, the clock will keep running – one more half hour at a time.

Equity paradox: promises and reality

The equity paradox
Core contradiction: Medical cannabis seeks clinical precision and legitimacy while recreational cannabis embraces cultural branding, yet both often occupy the same marketplace with conflicting priorities.

> "Despite the deep scars left by the war on drugs, they have demonstrated that cannabis can not only serve as a restorative plant for the body but also heal the soul of a community." – Social Equity in Cannabis [21]

The story of social equity in cannabis reform is one of aspiration meeting architecture. The laws are rich in intention – promising to repair the harms of prohibition through preferential licensing, record expungement, and community reinvestment – but the systems built to deliver those promises are often engineered for disappointment.

In state after state, equity initiatives arrived draped in moral language and bureaucratic fine print. The goals were noble: to return opportunity to the very communities most damaged by criminalization. Yet the implementation came with labyrinthine applications, high capital requirements, and a thicket of regulations that rewarded sophistication and seed money over sincerity and need.

As Garriott and Garcia-Fuerte observed in 2023 [22]:

"social equity provisions are now being included in most adult-use legalization schemes . . . [but] simply getting states to adopt such provisions may not be enough; the legal environment is proving hostile to social equity efforts."

Even when intentions are genuine, the market's architecture often works against its architects.

The result is a paradox worthy of its predecessors: programs designed to level the playing field often end up reproducing the very hierarchies they were meant to dismantle. Investors partner with equity licensees to gain access to permits, then buy them out. "Community reinvestment funds" are diverted to administrative costs or spent on police training. What the market needs to be viable – stability, scale, capitalization – is rarely compatible with what people need to thrive – access, trust, and time:

"What the market needs to be viable is not always compatible with what people need to thrive."
– Garriott and Garcia-Fuerte, 2023 [22]

Equity programs were meant to turn reparation into participation, but often they function as moral offsets for an industry built on unequal footing. For many who sought to enter the market, the promised redress of prohibition became its bureaucratic sequel: long waits, sudden denials, and another kind of exclusion dressed in progressive language.

This, too, echoes the pattern of the *Decriminalization Paradox*: reform achieved through definition without the instruments to measure fairness. Equity programs often lack the evaluative infrastructure to verify whether they work – no standardized data on ownership, no transparency in distribution, no feedback loops to show progress. The equity ideal exists in statute but not in spreadsheet.

Meanwhile, the cultural split between "medical" and "recreational" cannabis continues to fracture public understanding:

"Probably one of the oldest plants known to man, cannabis was cultivated for fiber, food and medicine thousands of years before it became the 'superstar' of the drug culture." – Vera Rubin, Cannabis and Culture [23]

The medical market aspires to stability – controlled cultivation, predictable dosing, measured outcomes. The recreational market, by contrast, thrives on novelty, branding, and the pursuit of sensation. Between them lies an unstable boundary where commerce meets care. Medical products are sold under names like Kush, Haze, or AK-47 – folklore turned pharmacology. To seasoned users these names are harmless affectations; to new patients they can sound like relics from an arms catalog:

> "While longtime cannabis users may think nothing of names like AK-47 or Alaskan Thunder F#*
> $, such monikers can . . . turn off people because they are offensive, sound scary or invoke
> deadly drugs." – All Natural MD, 2017 [24]

For clinicians, such branding undermines trust. For patients, it muddies the line between
treatment and recreation. The result is another quiet inequality: patients who must navigate a marketplace that speaks in code. Each new strain name adds to the noise, and the
signal – the product's actual chemical or therapeutic profile – grows fainter:

> "Every time you go to the dispensary there's 30 new names that you've never heard of," – Mara
> Gordon, co-founder of Aunt Zelda's Inc. as cited in Green Relief [25]

Genetic analysis confirms the intuition. In 2019, researchers found that strain names
assigned by cultivators bore little relation to genetic identity or chemotype [26]. The
abundance of labels disguised a paradoxical sameness: marketing diversity without
molecular difference. What appears to be a vibrant ecosystem of innovation often
masks a monoculture of mimicry.

Even the old folk taxonomy of sativa versus indica – still used by dispensaries,
marketers, and patients alike – has collapsed under scientific scrutiny. Studies reveal
no consistent genetic or chemical boundary between the two. The labels persist not
because they are accurate but because they are convenient, a marketing shorthand
that resists extinction:

> "The illusion of diversity fueling the Sativa vs. Indica debate is clear evidence suggesting that
> such labels are meaningless. There exists no significant chemical or genetic difference between
> the two." – Jahan Marcu, Rolling Stone, 2025 [27]

The Equity Paradox, then, is not limited to economics or access. It extends into language, branding, and knowledge itself. Equity cannot thrive in a system that confuses
identity for variety, that mistakes nomenclature for innovation. Without a common
vocabulary grounded in measurement and transparency, the dream of inclusion remains just that – a dream drafted into statute.

Technology and AI: promise and peril

The future paradox
Core contradiction: Technology and AI promise to solve cannabis information problems, yet they may
amplify existing biases.

> "Be nice or I'll replace you with a very tiny shell script." – Anonymous

It is an old human habit: to hand our uncertainties to the machines we've built – be
they abacuses, telescopes, or now the silicon prophets of artificial intelligence. Why

hire a junior analyst, a copywriter, or a clerk, when the machine will, for the price of a monthly subscription, conjure their labor in seconds? Blogs bloom overnight, market forecasts materialize at the click of a key, regulatory summaries materialize before coffee is brewed. The drudgery of thought – its drafts, errors, and hesitations – can be replaced by the glossy efficiency of code.

Cannabis, long trapped between stigma and speculation, has become an eager testing ground. Algorithms now parse terpenes, model consumer behavior, and design personalized dosing. The faith is familiar: that if only the data were large enough, the machine could cleanse the bias and the burden from human judgment. But the record is flawed at its very source. Research papers larded with sexism, policy memos soaked in racial bias, whole shelves of science warped by political expediency and funded to find specific outcomes – these are the waters in which neural networks drink. Feed them all the data, wind the turbines of computation, and you do not cleanse the errors – you automate them. Outsource our judgments to machines, and the result is not neutrality, but a more efficient prejudice.

The world is already awash in cannabis information, a digital Babel of forums, podcasts, blogs, and tweets. Into this glut, generative systems pour gasoline. They can spin up thousands of half-new voices each week – copies of copies – each echoing the same curated talking points, amplifying what was already distorted. The bar for participation has never been lower, and with that ease comes not enlightenment but acceleration of bias. And once the all-mighty algorithm knows you're interested in certain things, the targeted advertisements may never end:

> "I think there are a lot of people that won't allow themselves to use cannabis because of the stigma or the belief system that they're maintaining. That is probably a false narrative, not only in society, but for themselves so." – Asia Taber, 2021 as cited in Black [28]

The cannabis industry itself carries its own inherited gravity: stigma. Born from a century of demonization, it remains bound by the perception of cannabis as a taboo indulgence, unfit for polite commerce. Entrepreneurs find their progress hemmed in not just by regulations and hostile banks but by suspicion from partners who see "weed" rather than "enterprise." If stigma is a prison, the keys are education, advocacy, and policies anchored in evidence. Only then can cannabis take its place within the ordinary ecosystem of business, innovation, and investment.

Here, artificial intelligence is cast as both the peril and the promise. Scholars write of its utility – real-time crop monitoring, chemical analysis for safety, personalized dosing to reduce harm, forecasts of market shifts [29]. Others note AI's capacity to expand equity, to democratize access to knowledge, to aid entrepreneurs starved for data. AI, in its best light, can be telescope and microscope both: revealing unseen therapeutic potentials, predicting global patterns, sifting through genomes for new medicinal avenues. In this role it could help cannabis step from shadow into recognition, legitimacy, perhaps even respect:

"All models are wrong, but some are useful" – George Box [30]

But to imagine AI alone dissolving stigma is to misread stigma, machines, and our models for making predictions. Stigma is not an error in a spreadsheet; it is sediment layered over generations – social, cultural, political. AI can assist in the excavation, but it cannot exhume the truth unaided. It can lend transparency, offering cannabinoid profiles, potency maps, and consumer education that dispel myth. It can even accelerate drug discovery and validate medicinal claims long denied. Yet de-stigmatization is no single algorithm's triumph. It is the slow work of policy, of shifting norms, of culture catching up with evidence.

If there is a lesson here, it is the one often reminded us of: tools extend our reach, but never absolve us of responsibility. In the dance between cannabis and AI, we must decide whether the machines will amplify our errors or magnify our better angels. The future, as ever, depends less on the circuitry than on the choices of those who switch it on.

The political paradox

The political paradox
Core contradiction: Cannabis is the only substance whose form and strength are dictated not by doctors or chemists, but by politicians who neither study it nor use it, yet decide everything about it.

"I now have absolute proof that smoking even one marijuana cigarette is equal in brain damage to being on Bikini Island during an H-bomb blast." – Attributed to Ronald Reagan [31]

If politics is theater, cannabis has long been its most miscast actor. Few subjects have been so relentlessly rewritten to suit the script of the moment: menace, miracle, menace again. When the U.S. Food and Drug Administration, in a rare turn, approved THC (aka Marinol) to treat wasting conditions, and CBD to treat two rare pediatric seizure disorders, it thrust oncologists and pediatricians – not politicians and lobbyists – into the bright glare of controversy. Parents asked questions their doctors had not been trained to answer. The profession of medicine, long the custodian of empirical truth, found itself scrambling to interpret what was at once, both a scientific advance and a political act.

To call what's sold in dispensaries "medical cannabis" or "recreational cannabis" misses the point. The truer label might be "Political Cannabis." Every gram, every tincture, every pre-roll is the distilled essence of decades of congressional hearings, courtroom testimonies, and backroom deals. Unlike alcohol or tobacco – commodities consumed without euphemism – cannabis remains cloaked in adjectives. "Recreational" is absurd; society does not drink "recreational whiskey" or inhale "recreational Marlboros." "Medical" too often exceeds the evidence. What remains is a product more defined by political compromise than pharmacological clarity:

"Given the risks associated with marijuana, the nation needs the federal government to shift its posture from prohibition to regulation. To make that shift effectively, Congress needs to take a hands-on approach." – USA Today Editorial Board [32]

Do not be fooled. Congress has already taken a hands-on approach. That's how we got here. What remains is a product more defined by political compromise than pharmacological clarity. The next time you see someone draw on a vape and exhale a large synthetic cloud, remember that the cartridge fueling the device is not just plant extract suspended in a proprietary excipient – it is the distillate of committee hearings, lobbying budgets, and bureaucratic half-measures. Each puff is an aerosolized act of governance, a vapor trail of compromises and the cautious signature of local officials who wouldn't know a trichome from a chloroplast. The device in that hand may look like consumer tech, but in truth it's the latest delivery system for American politics – politics heated to over 400 degrees and inhaled straight into the lungs.

The myths shaped by politics persist with the tenacity of folklore. One such tale is that prohibition itself bred higher potency – the so-called Iron Law of Prohibition. The story goes that illegality compels concentration, pushing weaker versions off the market. But this too is a convenient fiction. Alcohol grew stronger after repeal, tobacco was fortified long under legality, and pharmaceuticals are routinely engineered for greater intensity within the open laboratory of regulation. Potency does not emerge from shadow; it grows wherever science and commerce converge. To blame illegality is to confuse correlation with causation, politics with chemistry. High-potency cannabis, so often paraded as a public menace in headlines, is less the child of prohibition than the handiwork of legislators themselves. The government wags its finger – beware the strong stuff – while the industry shrugs: it's the only thing we sell. The regulations and tax structure reward what can be quantified and monetized: the highest THC number on the label. Incentives align, the marketplace responds, and the result is an endless march toward stronger, narrower products. Consumers are left in a paradox: cautioned at every turn, yet offered no real alternatives:

"Medical science has made such tremendous progress that there is hardly a healthy human left." – Aldous Huxley as cited in Muldoon [33]

Underlying all this is a deeper problem: the scientific illiteracy of politics itself. No active scientist holds a seat in the U.S. Congress. Federal funding rules discourage researchers from entering races, and the structure of political life punishes those who cannot work campaigns full time. To run for office while running a lab is to court professional ruin – either give up your grant and your livelihood or limp along on part-time wages. The result is predictable: legislation written by those who have never sequenced a genome, never read a toxicology table, never stepped foot in a lab outside of a press tour. Laws about science are passed without science at the table.

In this vacuum, rhetoric metastasizes. One president compares a single joint to the fallout from a nuclear blast; another administration agonizes over scheduling cannabis in the same category as heroin while simultaneously patenting cannabinoids as neuroprotectants. The gap between scientific literacy and political decision-making is not a crack in the system; it is the system. Into that gap fall patients, physicians, entrepreneurs, and citizens alike.

Conclusion: transforming paradox into progress

The cannabis paradoxes mapped in this chapter reveal a society caught between the past and future, prohibition and acceptance, promise and reality. They are not merely contradictions of policy but mirrors of our collective temperament – our need to moralize what we cannot yet measure, to codify before we comprehend.

The quotations woven through this chapter illustrate the fragmented chorus of cannabis discourse. From Reagan's nuclear metaphors to Elders' calm reassurance, from the Surgeon General's contradictions to the frustrations of researchers and patients alike, the record speaks with many voices, often discordant. They are reminders that cannabis has never been allowed to stand on its own terms; it has always been spoken for – by politicians, physicians, journalists, advocates. To read these voices side by side is to witness not only the history of a plant, but the struggle of a society to define what it wants from science, medicine, and law.

Resolving these paradoxes will not come through rhetoric alone. It also requires a slow alignment across law, science, and culture: removing research barriers while developing new evidence frameworks, harmonizing federal and state laws while protecting worker rights, ensuring product safety while combating misinformation, and delivering genuine equity while building sustainable markets. Such efforts will not erase contradiction – paradox is endemic to human institutions – but we can channel it toward constructive ends.

Questions and exercises

Identify a cannabis policy and research what scientific information (if any) was cited during its development and implementation by regulators or policymakers.

Which research topics seem to get the most attention?

Compare cannabis policies between two different states – what paradoxes emerge from their different approaches?

Interview someone affected by workplace cannabis policies – how do they navigate the legal paradox?

References

[1]   Centers for Disease Control and Prevention. *Cannabis Facts and Stats*. 7 Mar. 2025, https://www.cdc.gov/cannabis/data-research/facts-stats/index.html.

[2]   Gallup. (2024). "What Percentage of Americans Smoke Marijuana?". Gallup. 1 Nov. news.gallup.com/poll/284135/percentage-americans-smoke-marijuana.aspx.

[3]   NCSL

[4]   Flowhub. (2025). 2025 Marijuana Industry Statistics & Data Insights. Flowhub. www.flowhub.com/cannabis-industry-statistics.Flowhub.

[5]   Last Prisoner Project. (2025). "The Unacceptable Reality of over 200,000 Cannabis Arrests in 2023.". Last Prisoner Project. www.lastprisonerproject.org/the-unacceptable-reality-of-over-200-000-cannabis-arrests-in-2023.lastprisonerproject.org.

[6]   NORML. (2025). *"DEA Seizes over 5M Cannabis Plants in 2024, Reports Nearly 6,000 Marijuana-Related Arrests*. NORML Blog. 4 Aug. norml.org/blog/2025/08/04/dea-seizes-over-five-million-cannabis-plants-in-2024-reports-nearly-6000-marijuana-related-arrests.

[7]   Ladouceur, R. (2018). The cannabis paradox. Can Fam Physician, 64(2), 86.

[8]   Carpenter, D.P. (2010). Reputation and Power: Organizational Image and Pharmaceutical Regulation at the FDA. Princeton University Press.

[9]   Roberts, C. (27 Feb. 2025). "Trump Research Cuts Threaten Cannabis Studies, Pose Rescheduling Questions." *MJBizDaily*, https://mjbizdaily.com/trump-research-cuts-threaten-cannabis-studies-poses-rescheduling-questions/

[10] Adlin, B. (2025). Marijuana's Schedule I Status 'Traps Researchers in a Paradox,' Federally Funded Scientists Say. Marijuana Moment, 3 July, www.marijuanamoment.net/marijuanas-schedule-i-status-traps-researchers-in-a-paradox-federally-funded-scientists-say/.

[11] Titus, D. (29 Apr. 2025). "Reps. Titus, Omar Introduce Evidence-Based Drug Policy Act." *Office of Congresswoman Dina Titus*, https://titus.house.gov/news/documentsingle.aspx?DocumentID=4764.

[12] Grinspoon, P.M.D. (2020). "Medical Marijuana." *Harvard Health Publishing Blog*. 10 Apr. www.health.harvard.edu/blog/medical-marijuana-2018011513085.

[13] Koop, C.E. (1982). Surgeon General's Advisory on Marijuana (Draft). U.S. National Library of Medicine Digital Collections. collections.nlm.nih.gov/catalog/nlm:nlmuid-101584930X675-doc.

[14] Adali, L. (22 Feb. 2013). *"Medical Marijuana Uncovered!" Nature Education: MedSci Discoveries*, www.nature.com/scitable/blog/medsci-discoveries/medical_marijuana_uncovered/.

[15] CNN Newsource. (2023). Study: Marijuana Users Have More Heavy Metals in Their Bodies. CNN Health, 30 Aug. www.cnn.com/2023/08/30/health/marijuana-heavy-metals-wellness/index.html.

[16] Kleidon, A.M., Peterson, A.M., Warner-Maron, I., Glicksman, A. (2023). Attitudes, Beliefs, and Perceptions on Cannabis Among Older Adults Aged 65 and Older: A Cross-Sectional Survey. Journal of Primary Care & Community Health, 14, article 21501319231177284. https://doi.org/10.1177/21501319231177284.

[17] Müller, M., Mészáros, E.P., Walter, M. et al. (2023). Cannabis Consumers' View of Regulated Access to Recreational Cannabis: A Multisite Survey in Switzerland. European Addiction Research, 29(3), 213–221. DOI: 10.1159/000530194.

[18] Noguchi, Y. (5 Feb. 2025). "Vaping Weed Is Very Popular, but Users Should Be Aware It Carries Risks." *NPR*, www.npr.org/2025/02/03/nx-s1-5246302/vape-marijuana-weed-pesticides-chemicals.

[19] Wozniak, E. (10 Mar. 2025). "The Hidden Risks of Legal Cannabis: How State-by-State Regulations Contribute to Contamination and Health Hazards." *UC Law Review Blog*, uclawreview.org/2025/03/10/the-hidden-risks-of-legal-cannabis-how-state-by-state-regulations-contribute-to-contamination-and-health-hazards/

[20] National Association of Attorneys General. (24 Oct. 2025). *"39 State and Territory Attorneys General Call for Clarification of Federal 'Hemp' Definition."* Press release, https://www.naag.org/press-releases

/bipartisan-coalition-of-39-state-and-territory-attorneys-general-urges-clarify-federal-definition-of-hemp/. Accessed 24 Nov. 2025.

[21] Roots, S.F. *"Social Equity in Cannabis."* SFRoots, https://sfroots.com/social-equity. (Accessed 18 Nov. 2025)

[22] Garriott, W., Garcia-Fuerte, J. (2023). The Social Equity Paradigm: The Quest for Justice in Cannabis Legalization. Seton Hall Journal of Legislation and Public Policy, 47(Iss. 1), Article 4. Available at https://scholarship.shu.edu/shlj/vol47/iss1/4.

[23] Rubin, V. (ed.). (1975). Cannabis and Culture. De Gruyter Brill. Chapter 1, "Introduction," 1–12, or whatever the page range is. DOI: 10.1515/9783110812060.

[24] All Natural. (2017). "Medical Marijuana Strain Names Are Offensive." *Florida Medical Marijuana Blog.* 11 Oct. floridasmedicalmarijuana.com/blog/medical-marijuana-strain-names-offensive/.

[25] Green Relief NZ Ltd (GREENLAB). (2020). Cannabis Strain Names Are Meaningless: What Is the Industry Doing About It? Greenlab, greenlab.co.nz/cannabis-strain-names-are-meaningless-what-is-the-industry-doing-about-it/.

[26] Schwabe, A.L., McGlaughlin, M.E. (2019). Genetic tools weed out misconceptions of strain reliability in Cannabis sativa: Implications for a budding industry. Journal of Cannabis Research, 1(1), 3. https://doi.org/10.1186/s42238-019-0001-1.

[27] Marcu, J. Ph.D. (29 Jan. 2025). Waking from the Strain Dream: Cannabis Science at a Crossroads. Rolling Stone Culture Council, www.rollingstone.com/culture-council/articles/waking-from-strain-dream-cannabis-science-crossroads-1235248871/.

[28] Black, C.M. (1 Nov 2021). Buds, Bones & Broken Stigma. Cannabis Culture, www.cannabisculture.com/content/2021/11/01/buds-bones-broken-stigma/.

[29] Weinberg, C. (2024). Revolutionizing the Cannabis Industry: In the AI Era. *2024 Portland Int.* Conference on Management, Engineering and Technology (PICMET), 00, 1–5.

[30] Wicklin, R. (2 Apr. 2025). Did George Box Say, 'All Models are Wrong, but Some are Useful'? In: The DO Loop. SAS Institute, blogs.sas.com/content/iml/2025/04/02/all-models-are-wrong.html.

[31] Sreechinth, C. (2019). Ronald Reagan's Legacy of Words: 1000+ Quotes of Ronald Reagan. UB Tech, Google Books. books.google.com/books/edition/Ronald_Reagan_s_Legacy_of_Words/3-oEEAAAQBAJ?hl=en&gbpv=1&bsq=marijuana.

[32] *USA Today* Editorial Board. (2022). Time for Change: Federal Ban on Marijuana Use Causes More Harm than Good. usatoday.com. July 31.

[33] Muldoon, L. (2014). Pandemic of great expectations. Can Fam Physician, 60(9), 797.

Experts in the field interview with Michael Loconto, Esq.

Michael Loconto is an attorney and alternative dispute resolution provider who serves as an independent arbitrator and mediator in workplace, commercial, and consumer disputes through his nationwide practice, Loconto ADR, as well as through the American Arbitration Association's labor, employment, pension/ERISA, and consumer rosters. He is also available via the Labor Relations Connection, FMCS, and numerous state and local panels, and hears rail carrier disputes for the federal Surface Transportation Board. In addition to supervising union organizing and certification processes and conducting bargaining training, he has trained with the FMCS Institute, apprenticed with members of the National Academy of Arbitrators, and participates in the NAA New England arbitrator salon. Over two decades, he has built deep experience across higher education, K-12, construction, public safety, dining, athletics, trades, transportation, and tech, addressing issues such as discipline and discharge, contract interpretation, benefits, discrimination, disability, and family and medical leave. He previously served as general counsel for a college, a labor and compliance leader at a university, and a labor attorney for a major city, while also teaching legal writing and research at Northeastern University School of Law and bargaining and contracts at The Labor Guild School of Labor-Management Relations. Active in the field, he co-led the LERA Higher Education Industry Council, serves as President of the Boston LERA chapter, and frequently presents on labor, employment, and education issues.

Jahan Marcu

Hello, I am Dr Jahan Marcu

Andrew Peterson

And I'm Dr Andrew Peterson, and we are the editors of the *Cannabis Innovation Series*. And today we welcome another interviewee to our experts in the field. This second volume of our series is focused on *Demystifying Cannabis*, where we examine myths, mysteries and truths about various topics in the industry. We're here today to talk about drug testing and cannabis. With us today is Michael Loconto, an arbitrator and mediator in the Boston area.

Jahan Marcu

It is interesting to talk with someone who works with data and expert witnesses as it relates to cannabis, particularly since there is so much controversy and evidence within the industry and it becomes difficult for people and organizations to develop policies anchored in truth. So tell us a little bit about what you do in this area so the listener can better understand your role, and if you can address how you work with expert witnesses or subject matter experts in your field?

Michael Loconto

Well, certainly. Dr Marcu, Dr Peterson, good to see you today, and pleasure to be here. Thank you for having me. So as you mentioned, I'm an arbitrator and a mediator in the labor and employment field, and so where most often these issues come before me are in a hearing setting, the type of hearing that you have in an arbitration and I'll focus on arbitration, because that's the bulk of what I do, is a lot like what you find in a courtroom, and I act a lot like what you would see as a judge, except, as you can see, I don't wear robes and wear a wig. You know, give me a few of the drugs we're going to talk about, and things might change.

Nevertheless, we have a practice that looks a lot like what people see on "Law & Order", except translated into the civil field. We have lawyers, we have objections, we have sworn testimony. And the idea is that we're getting to the heart of the matter with respect to what happened in a given instance. Typically where you see drugs in these disputes is you've got a union member, so when we're talking about arbitration, we're talking about a dispute between an employer and a union, it's a dispute that's grounded in the contract, the collective bargaining agreement, between the parties. And in that collective bargaining agreement, you have what's called just cause for discharge or discipline. And what that means is that the employer has to have just cause for terminating or imposing discipline on a member of the Union. And that's a very highly developed concept in our field, and where it relates to drug testing is often when you have somebody that's been terminated for being intoxicated on the job, we have testing that ensues. Often, we might have some field sobriety testing that takes place, and in those instances, we also have a very highly developed practice where we need expert witnesses to walk us through. I am not a scientist by trade. I'm a lawyer by trade. I always like to tell the joke that my father thought I was going to be a NASA engineer because the science teacher told him I was "taking up space". But having said that, I'm willing to learn, and that's the attitude that you have to take when you have experts in the room is you have to have a willingness to learn from individuals who are qualified to give you their opinion, and their opinion, more importantly, as it relates to the facts of the substance that's at issue, the facts of the case, and how that relates, ultimately, back to the contract between the parties. I hope that was a very boiled down introduction on a very wide area of work. But let's see where we can go from there.

Andrew Peterson

So Michael, I find it interesting you're bringing in the subject matter experts to kind of give the "straight poop" about, you know, drug use and drug testing and things like that. But there's got to be some misconceptions out there that these experts are coming in and kind of clearing up. What are some of those misconceptions that about drug testing that you wind up seeing or hearing?

Michael Loconto

Certainly, I would say probably the biggest issue that that's bubbling up in our field today is that you have some level of certainty out there about how you can demonstrate that an individual is actually intoxicated by marijuana. For instance, marijuana is a big issue right now because the legalization and the changing effects of legislation and new laws around the country on the workforce and what people think they can do at work or prior to coming to work, and the quick answer is, there's no reliable source to demonstrate intoxication by cannabis. And we have experts that might come and talk to you a little bit about maybe saliva testing. For instance, I'm sure both of you are familiar with that, but the last statistics I saw were that was that saliva testing, even at the best model, is maybe only in the 70s or 80s. So if you only want to get it right four out of five times, then I guess you can go with saliva. But we're more about the truth, and we're more about trying to find whether an individual is actually broken workplace rule.

Andrew Peterson

So when you say that there's no kind of hard and fast rule about intoxication with marijuana, like with alcohol, we've got a blood alcohol concentration – many states at 0.08 – is that what you're talking about in terms of that hard and fast rule?

Michael Loconto

Yeah, you know alcohol cases. The law is so well developed around alcohol that those cases typically don't come forward very often. A classic example of intoxication by alcohol in the workplace might come more from a less sophisticated workplace, if you will, where you've got individuals that may or may not be trained in how to spot the effects of intoxication. Viewing an individual acting erratically at their workstation, they pull them off the line, they talk to them. You know, all the indicators of just being a human being and interacting with people is telling them this person is intoxicated. However, when it comes to making a determination about whether that actually is the case and whether that violated any workplace rules that defense attorneys, or in this case, union advocates, have a number of different avenues of attacking that finding. You know, how were the individuals trained to spot the effects of intoxication? Did you send the person for a test, which typically is not going to be the case in alcohol because it leaves the system so quickly? Did you have any other sort of fail safes that you could use to perform? Excuse me, to test whether the individual was intoxicated, and did any investigation follow that may have bolstered your suspicions about the individual? So, all of this stuff comes out through a number of different methods in the hearing. We're talking a little bit about drug testing experts right now, but you might have investigatory experts. You might have individuals that can speak to the practice of the workplace and so on and so forth. And all of that is bound up in what the practices have been within that particular workplace, or the industry more broadly, or in our field around intoxication due to drugs and alcohol.

Jahan Marcu

You bring up a lot of interesting aspects of this that you know, employers or organizations might have the best of intentions of creating a great work environment, but some of the, I guess, pragmatic or practical implementations of that policy you know, might present a challenge. So how might drug testing policies unintentionally reinforce systematic inequities or just barriers to employment that might be maybe unnecessary or a little over the top, or just some I guess I would say, unintentional consequences of drug testing policies?

Michael Loconto

Well, that's a great question, and there are a lot of there are a number of industries in particular that have met this issue head on. I've worked a lot with the construction industry over my career, and in a number of trades, skilled and unskilled. You find a pretty advanced, and I guess I'll call it pragmatic policy that has been developed over the years where typically having either a conviction in the criminal sense or a termination for drug-related offenses is an absolute bar to employment on a worksite. Practically speaking, the construction industry has determined that that actually wipes out a whole swath of workers that might be available to us. And so what's the next alternative? – Rehabilitation.

And you know, I'll call out the carpenters Union, in particular, they have an extremely robust rehabilitative process for individuals who have, for example, maybe a DUI or something like that that might cause a barrier, but people that have been hooked on, you know, really, really tough stuff that that it's hard to break free from. I'm talking like, you know, Percocets, from on the job injuries and things of that nature. So, I don't want to be all doom and gloom. I started with that because I wanted to highlight responses to some of the systemic inequities out there. I want to, you know, shine some light on good work that's happening. But at the same time, not all industries, not all employers, are as enlightened as that. And you can have a lot of individuals who, due to a background screen in a criminal record check, they might be picked off because of a DUI for a position that requires you to operate a company vehicle. There might be a checking of references that reinforces that the individual had some trouble at a previous employer, and that can really create some of those systemic inequities that you speak of with respect to initial employment. On the other side of the coin, once individuals are in in the workplace, drug testing policies can be 100 pages long in some instances, and that really can pose a problem, even for the HR people that are that are called upon to implement those policies, much less, the average Joe working on the line, and maybe their union representatives as well. So there's a lot of opportunity there for, I think, individuals to have the system used against them. And that's what the process that I'm involved in is intended to, to suss out. We are a justice based process, and that just cause standard that I talked about earlier within disciplinary and discharge matters is intended to determine whether a rule exists, whether it's fair, whether an employee had notice of it, whether they had an opportunity to correct their behavior prior to taking additional discipline – things of that nature that all go to the process, as opposed to the violation itself.

Jahan Marcu
Yeah, I had not thought about the impact that drug testing policies would have on HR. That is, I could see how, if you're at that level and HR is trying to resolve these issues, to implement them fairly and objectively, that if they don't understand it, that could be an issue for an organization as well. Anyway, that's just a pondering I had from what you said.

Andrew Peterson
Related to that. Michael, what recommendations do you have for an employer who's trying to create some drug policy testing that makes it both fair and effective, and especially in light of like marijuana, where there's no kind of cut off, or we're still in that whole, discovery area? What about recommendations?

Michael Loconto
So you know, first of all, every policy has to fit the workplace. You know, every workplace is different, so I wouldn't take something off the shelf and just implement it. You've got to modify it so that it fits your needs and the practicalities of your workforce. But having said that, if you can recreate the wheel with something that works with another employer in your industry, then I'd say that's a great starting point. And you know, I call out the carpenters as one example for construction. But every industry has its exemplars, and that's really in my field, the benefit of having a collaborative relationship with your union counterparts on the other side of the table.

I worked in management before I was a neutral, an arbitrator, and every policy that I tried to implement through collaboration and starting from the beginning in a collaborative process for drafting and implementation worked out much better than anything that I imposed upon the people across the table. You have a chance to break down some preconceived notions that we've talked about earlier. You have a chance to talk through perceived pain points that maybe you anticipated the outset, or maybe as you talk it through with your colleagues, you determine a light bulb goes on and says, geez, I didn't think about it that way. Maybe we can change this so that both of us can achieve our objectives. That type of time that you put in at the outset, can really help reduce the type of friction or simple mistakes that might happen later on. And then you have to train. You know that's training is so key, particularly in a science based field like drug testing, where most of us are allergic to science, right? Or at least, you know, I don't want to say ignorant. I'll speak for myself a little bit from time to time, maybe just because that's not been my focus of my education over the years. But certainly, many people don't fully understand what cutoff levels mean, what the different classifications of drugs are, how they relate to criminal laws, how they relate to the type of behaviors that you can view in the workplace and how it may negatively affect worker output. All of those things are things that you can't just we talked about an imposition on the other side of the table, but management can't expect to impose that on the people on its own side who are intended to enforce that. They've got to have some training to understand a few baseline concepts.

Andrew Peterson

So let me get this right, then don't take something off the shelf, but it's okay to find something within the industry, but tailor it to your industry, and particularly in collaboration with the partners who are going to be affected by it. And then once that's done, you need to make sure that people are trained on the policies themselves, and also in in the idea of detection and the like. If you do those things, that would help to create a better employer drug testing policy.

Michael Loconto

Yeah, you've exceeded the work of about 95% of the parties out there if you've done those simple things.

Jahan Marcu

It all seems like, seems like there's so many factors to consider. I mean, are you a state that has adult use or just medical will you make exceptions? What do you do if someone takes a hemp product that has trace amounts of THC, and so I think about all these factors, and I gotta wonder if there are resources you might recommend that people check out. Is there a book or podcast that you recommend or you find yourself sharing frequently? I know there's certain videos and things on the endocannabinoid system and aspects of cannabis I share frequently with people, versus having a three hour conversation, I'm like, watch this short cartoon. But maybe there's something else out there. A couple of resources you'd recommend for folks.

Michael Loconto

Yeah, well, it's great you asked that, because with respect to cannabinoids, there's really not a lot out there that connects to the workplace. We're working to solve that, and Dr. Marcu, you're helping along with one of those projects, which will be published next year through the Labor Relations Information Systems (LIRS) Press out in Oregon. It's a book on drug testing in the workplace, and we're going to include a number of updated chapters relating to cannabinoids. And I say that as an update, because really, the next best workplace resource on drug testing was published in 1994. Now, looking back, folks might realize there's been a few changes in the drug testing landscape and the drug legalization landscape in this country, and particularly as it relates to workplaces in the last 30 years. And so, we're making an effort to make that a little bit more accessible in everyday language for folks that need to interpret these policies, implement these policies, run the whole gamut. Having said that, if you want to get a baseline that original book, Drug Testing Issues and Arbitration by the Denenbergs' was published by BNA books in 1994. Folks can find it. It's should still be in circulation in most law libraries and the like.

On a training side, there are a number of trainings out there that law firms provide. LRIS has a training component to it as well. And I believe there's a cannabis webinar that's available on their website for download that that's pretty informative on the state of the art with respect to laws and how they relate to the workplace. But, you know, generally speaking, what I would tell people from a practical standpoint is, if you're an attorney and you work in this field, just go out there and find yourself a few arbitration decisions that have been published that relate to drug testing. Anyone can do a Lexis or Westlaw search for drug testing cases. And, you know, they're not going to tell you what to do, but they're sure as heck going to tell you what not to do. Because that's typically the type of cases that make it through to decision and publishing. And I say that because, you know, I feel like I'm painting myself as kind of, you know, the village idiot here, right? When I say, look, I don't know much about science, but I always say, the but is what matters. I'm willing to learn. When you look at a lot of the decisions that are out there, you will see that there are often opportunities for individuals like myself, in the neutral role, as well as parties on each side, to engage in a little bit of that learning that might have saved a whole lot of time and money and avoided that litigation process and resulted in an adverse decision for one side or the other. And, you know, learning about, you know, what are some of the qualifications of the MROS (medical resource officers) that come before individuals like myself, who you usually see in a drug testing context, and you know, what are their qualifications, and what do they know about the substance and the testing process that's at issue, what are some of the things that employers relying on when they take a case forward? How does a union effectively defend against a charge? All of those things can be very instructive, and I find that, you know, it's like the old saying goes, I learned much more from my mistakes than from my successes. And there's a whole host of resources out there that folks can take advantage of to do the same thing.

Andrew Peterson
That's helpful, really helpful. I appreciate that, and we'll do the best that we can to identify a couple of these resources for our listeners and readers as we move through this. So thank you very much for sharing that information and answering our questions today. You've been very helpful, Mr Loconto. This has been informative and enjoyable to learn a little bit more about the complexities of drug testing and policies in the workplace and how people like you help to continue to pave the way for an improved system for fair and equitable drug testing policies.

Michael Loconto
Well, thank you. Dr Peterson, it's a pleasure to be here, and I appreciate the opportunity. And most of all, thank the two of you for doing this work, you and your colleagues to make this type of information accessible.

Andrew Peterson
Thank you very much.

Jahan Marcu
Thank you.

TRUTHS

Chris Conrad

Who gets to be an expert? Cannabis, the courts, and credibility

Introduction

Anybody can call themselves a cannabis expert. However, when people end up in criminal courts or fighting a civil case, they need a particular kind of expert: A court-qualified cannabis expert witness [1]. Typically, such an expert serves as a consulting expert or a testifying expert. The work is fascinating, often complex, and the stakes can be enormous, whether it is the freedom of an individual facing prison time or the resolution of a civil dispute, sometimes involving millions of dollars.

People who have training through an institution like Oaksterdam University[1] or have done course work at a state-accredited institution have valid claim to the academic training that demonstrates expertise in a selected area of cannabis [2, 3]. Preferably, one also has specialized training by having worked as an employee or a paid consultant to generate professional experience. An expert should keep up to date on the legal and research developments in their field of expertise and there are scientific, legal,[2] and trade[3] conferences to do just that.

An expert witness is different from a percipient witness in two ways: A percipient witness had some direct observation of, or involvement in, the incident, such as an eyewitness to an accident or crime or an attending physician. Percipient witnesses are allowed to discuss only matters of which they have a direct first-hand role, observation or knowledge, such as a custodian of records. An expert witness is not involved in the incident and relies on other people's observations, photos, documents, or testimony to provide neutral evaluations and offer an informed opinion. An expert may rely on second-hand information, or hearsay, to draw relevant opinions from the evidence and express them in court (see Box 9.1). In some instances, it may be appropriate for an expert to conduct some sort of approved analysis or prepare a courtroom display on evidence or documents.

---

1 Oaksterdam University, "Training for the Cannabis Industry," was the first such academic institution, founded in 2007 by cannabis entrepreneur Richard Lee. Chris Conrad taught the history class until the Oakland CA campus was closed due to the COVID-19 pandemic. Its curriculum continues online. http://oaksterdamuniversity.com.

2 For example, NORML legal committee CLE trainings. https://norml.org/lawyers/legal-seminars/.

3 For example, National Cannabis Industries Association, http://theCannabisIndustry.org/.

© 2026 Walter de Gruyter GmbH, Berlin | https://doi.org/10.1515/9783111476162-011

> **Box 9.1:** Core activities performed by technical and scientific experts when assisting in investigations and legal proceedings.
>
> **A consulting expert may:**
> - Analyze reports from police and other experts
> - Evaluate photos and recordings
> - Review data, facts, and figures
> - Perform relevant experiments
> - Produce a report or declaration
> - Examine site/physical evidence
> - Suggest questions for one's self and opposing experts
> - Provide technical context

The role of a cannabis expert

Different courts have different jurisdictions, such as federal criminal, state criminal, federal civil, state civil, municipal courts, and administrative hearings. A court-qualified expert witness brings specialized knowledge of a specific topic to clarify a relevant aspect of a case for the trier of fact (judge, magistrate, or jury). The designation is rendered at a trial or evidentiary hearing. It means you've been on the stand and accepted as an expert by a sitting judge who rules for the record if you are qualified, on which topics you may testify, and what is relevant to the case.

Experts offer consultations, evaluations, reports, testimony, and depositions for attorneys, law firms, judges, juries, government agencies, and insurance companies in state and federal court hearings, trials, administrative hearings, and arbitrations. After you consult, if the attorney's legal theory and argument are not sound, that is usually the end of it. Otherwise, you may need to do a field investigation or write a report to be used to negotiate a reasonable settlement. In event of a conviction or lost ruling, an expert can still consult for appeals, provide affidavits or favorable testimony regarding sentencing and probation issues by reviewing mitigating circumstances or disputing the severity of the behavior.

Being a qualified as an expert doesn't necessarily mean you will end up on a witness stand. To the contrary, most cases settle before they reach that point. To testify, an attorney must ask questions about your qualifications before a judge, so write up some questions that showcase your resume. Opposing counsel can challenge your expertise or stipulate to it and attack you on cross examination. A judge may ask for additional information before making a decision. Each jurisdiction has its own rules but similar considerations, for an example see California Evidence Code §720 [4]. While courts in different jurisdictions may use different legal standards, in general, federal judges rely on the Daubert Standard [5]. This rule is intended to ensure that expert scientific testimony in federal courts is based on sound and widely accepted principles, methodology, and reasoning.

Based on my experience, barely one case in ten is likely to involve testimony or deposition under oath. A deposition is cross examination under oath, with a stenographer but outside the courtroom. You can correct a deposition, fixing a word or name, but you cannot rephrase or add new content. If you misspoke, you may need to correct it for the record with an affidavit or testimony.

Academic degree and life experience

Having a relevant postgraduate degree is the typical gateway to testify. Once you tell a judge that you have a degree in, say, cannabinoid research and/or published a book or research study, they are likely to accept you as an expert for the limited purposes of that one area. For example, I published the book *Cannabis Yields and Dosage*, which has been presented at juried conferences, updated several times and cited by others as an authoritative source [6]. Likewise an MBA helps you testify on business profits and losses, an engineering degree helps for indoor cultivation buildouts and energy use, and a medical degree will allow testimony in the field of medical use. You could then shop your publication or skillset around by joining a trade group[4] or sending a *curriculum vitae* to law firms that work in your area of expertise.

The problem with this route is twofold: First, academics may lack the *vitae* part of your CV – the real-world, lifetime experiences that a court requires for a nuanced parsing of legalities. Second, you may not have the right personality to handle hostile depositions, cross-examinations, and courtroom drama.

To the first point, a court may find someone who took a class on extraction to be less compelling than someone who took that same class as well as worked in the industry, dealt with technical problems, has experienced various scenarios and knows how to mitigate losses.

To the second point, confident humility is required. An expert witness must be circumspect, thick skinned and quick witted. A common misconception of a professional expert is to assume that your history will be treated respectfully by opposing counsel. To the contrary, they will do everything in their power to blindside you, smear your reputation and ridicule you. I've seen MDs take offense, quarrel, or get thrown off track by a hostile attorney. Appearing arrogant and argumentative doesn't help your credibility (see Box 9.2).

---

4 For example, Forensic Expert Witness Association, https://forensic.org.

> **Box 9.2:** Characteristics that enhance credibility and effectiveness of experts when presenting evidence and testimony.
>
> **What an expert brings to the table:**
> - Credible educational background
> - Specialized, real-life experience
> - Experience on the stand
> - Professional demeanor
> - Understanding of the law
> - Solid reputation
> - Engaging personality
> - Simplifies complex issues

Real-world experience has its drawbacks, as well. Once you see or know something you can't ignore it. Attorneys seeking to quash a search warrant may ask me to testify that vegetative cannabis plants (i.e., in the nonflowering stage of growth) have no distinctive odor. Based on experience, I can't say that, but I can say that's often true, and cannabis' vegetative odor is weaker, less pungent, less identifiable, more generic, and more prone to error than its flowers. Cannabis odor is a subtle interaction of terpenes [7] and thiol compounds, but cannabinoids do not emit any odor [8]. Hence, you can't judge potency, distinguish hemp from medical marijuana[5] or determine a specific weight based on odor alone [9].

Another example: Plant biomass is dried, manicured, and processed [10] to make usable marijuana.[6] For decades, law enforcement has trained police that every plant yields a pound of marijuana: "Each plant yields one pound of processed bud, according to my DEA training." When asked, however, they admit that their own field experience shows that's not correct. Agents frequently uproot plants that weigh only a few ounces fresh, including moisture, stalk, leaf, and maybe roots and dirt. The federal report *Cannabis Yields* [11] duly notes that one-pound estimate – but it also documents a 1990–1992 field study that discusses drying and manicuring. That study also provides both weight and canopy-based mathematical formulae to assess outdoor bud yields and the government has not disputed it. This results in very consistent testimony within the context of plant sex, lighting, fertilizers, pests, and the vagaries of agriculture.[7]

Attorneys are the gatekeepers for communications between client and expert. With the client's consent, the attorney hires an expert, who works for them with the

---

5 Cannabis plants containing less than 0.3% THC are considered either industrial or cannabinoid hemp under the federal Farm Bills passed in 2014 and 2018. https://congress.gov/bill/115th-congress/house-bill/5485.

6 With some exceptions, such as juicing raw cannabis or freezing before extraction.

7 See assorted books by Jorge Cervantes, Robert Clarke, Ed Rosenthal, or any other reputable cultivation author.

privilege of attorney confidentiality – but not client confidentiality. If opposing counsel asks you under oath if the client said something, you must answer. An expert objectively reviews facts and case theories, but forms their own opinion as to how it lines up with the evidence. Speaking to other experts and percipient witnesses is relatively safe.

Attorney, client, and expert relations

Sometimes, however, you need to interview a client. This presents a potentially compromising situation. Get the attorney's permission and, if possible, do so in the presence of counsel. Always explain to the client that you do not have confidentiality for these discussions. This is a very risky trade-off. I once asked to speak to counsel privately about seven possible defense arguments based on the discovery. The client insisted on participating, against his attorney's advice regarding the lack of privilege. Every time I suggested something innocent supported by the record, the client would exclaim, "That's not what happened," or "I wasn't doing that!" By the time I testified, I couldn't rely on any of those issues due to his comments. The client suggested instead, "just tell them nobody is stupid enough to do this stuff." Of course, the cop who testified before me humbly stated, "we don't catch every criminal, mostly just the stupid ones."

Before you do a client interview, therefore, tell the attorney what you will need to know, get approval for your questions and consider all the possible ramifications before you ask. Once they respond, you can get clarification but cannot ignore what they said. Statements you attribute to your client will be attacked as "self-serving" or may be stricken from the record unless the client takes the stand to authenticate their statements. That opens another can of worms.

Discovery, evidence, and investigation

One thing you do not want is to be on the stand with incomplete, inaccurate or biased information that falls apart and discredits you during testimony, especially during cross examination, when it may appear you were obfuscating or hiding information. Know the case's weaknesses and plan in advance how to respond. That's why you need to be completely honest and circumspect with retaining counsel.

As part of your case review, you may have to pore over large amounts of discovery and only a handful are useful to explain your case to attorneys, a magistrate, judge, jury, or insurance adjuster. Documents that were not submitted in court and been authenticated are not yet "evidence" but they are still subject to the rules of discovery and disclosure [12]. The federal Department of Justice provides guidelines for prosecutors to use regarding discovery [13]. Reciprocal discovery may require that your written communications and report be shown to opposing counsel, depending on its status as work product [14]. Phone calls are safer for highly confidential conversations

The bulk of criminal court discovery comprises police reports, legal documents, lab reports, utility bills, receipts, cell phone chats and data, logs, receipts, photos, audio and video recordings, and business records related to some event. You may be asked to interpret a chemical assay, chromatography chart, or blood test results. You need to digest this information, interpret it through the lens of your expertise, and anticipate issues that may arise, including opposing expert testimony. You may be called on to generate a report or sign an affidavit to proffer details about what you would testify to, should you be called. In civil court, you might need to discuss how a problem could have been anticipated or mitigated to reduce financial losses. Civil cases may require a deposition before opposing counsel.

You also may need to evaluate physical evidence that is in police custody. The attorney must arrange this and typically accompanies you on the inspection, as do opposing counsel, law enforcement and investigators. The court order should be detailed about everything you need to do: Visual inspection, taking measurements, weighing, and sometimes directly handling items. Property clerks are often nice and accommodating, but not always. I once went to a property room and the staff brought out several cardboard cartons. My request to open the boxes and look inside was denied. I asked them to weigh the boxes on their scale and they refused. I wrote a report saying they showed me no visible evidence of marijuana and defense counsel filed a motion to dismiss. The judge made sure I got to inspect the property before the next court date.

Your evidence inspection may or may not be recorded, but you need to make notes and take photos. If amounts are important, calibrate your scale using a physical weight. I use six nickels, which comes to 30 g. Make notes about dates, your client, calibration, tare weights, gross, and net weights, and how everything was calculated. Packaging is important, as are signs of pests and disease that affect quality and commercial utility [15]. You may request that certain items be tested for purity or potency (see Box 9.3).

Box 9.3: Common strategies and methods experts may use to interpret, contextualize, and communicate evidence in legal proceedings.

On the stand, an expert can:
- Corroborate other testimony
- Dispute interpretation of facts
- Discuss fact-based hypotheticals
- Rely on certain hearsay evidence
- Establish if evidence is consistent with common practices and behavior patterns
- Offer alternative explanations to interpret events
- Explain exculpatory circumstances
- Prepare charts, diagrams, and displays
- Challenge indicia of specific intent
- Illustrate ambiguity and reasonable doubt

Reading the courtroom

As an expert witness, the scope of your testimony is limited to what a judge decides is relevant. You can only consider information that is already in the court record to render an opinion. An attorney can ask hypothetical questions based on the evidence. Two benefits of being an expert are that you can consider hearsay and consult with other experts, although you may need to disclose or document such conversations at an evidentiary hearing or in an affidavit or deposition. Knowing a prosecutor's *modus operandi* is helpful when preparing for testimony.

The internal politics and personalities of a community and courthouse can be important. Respect the courtroom staff. Some judges have a relaxed court room with friendly clerks and bailiffs, while others will not tolerate the least perceived indiscretion, even if it's inadvertent. Try to get to know the clerk and bailiff. It may help to speak with local attorneys in a courthouse to get their insights about the district attorney (DA), judges, and local jury pool. If they like you, it makes everything go a lot easier. Wait to hear a full question, don't anticipate what is coming. Keep an eye on the court reporter. If you're like me, you can get wound up and talk very fast about a topic that you are very familiar with – like your area of expertise. Spell any technical words and people's names. If the stenographer looks stressed, pause to let them catch up. The more clearly and slowly you speak, the better the court transcript will be when jurors review it or someone orders a copy for the record or for an appeal.

One northern California prosecutor liked to burn up a significant amount of court time staring at me while formulating questions, trying to make me feel uncomfortable or force me back into court the next morning instead of finishing my testimony in one day. He carried a stack of court transcripts into court when I testified, looking officiously through the pages to pick cherries of confusion from unrelated cases. He would confront me with a transcript, light on a page, read a statement attributed to me and ask me to respond. I might reply with, "That sounds like something I might say," but I knew he might paraphrase me or misrepresent my implication, so always asked to review the document. Reviewing the context just before and after he stopped reading usually explained why I had made the remarks in question.

That same prosecutor transferred to eastern California, where I had many more cases with him but, ultimately, his courtroom tactics became predictable and his slow yet snarky approach to asking questions and attacking me would often turn off the judge or jury. Eventually, he became vindictive and made it a habit to slowly make a lot of objections during direct testimony and drag out his cross, so I would be held over for an extra day in court. He knew that I charged a flat rate, so that extra court day cost me money. From calendar conferences, he knew if there was a conflict with a different courtroom and forced me to reschedule other cases. Of course, I would never give him the satisfaction of seeing my irritation, so I remained cheerful about changes – but asked the attorneys set my testimony in the morning so he couldn't keep me overnight.

Choosing the right words

Have you ever used a colorful figure of speech and instantly regretted it? Whether talking to friends or testifying in court, getting the words right is critical. A good expert witness knows when to be detailed, when to be brief and when to be vague. In court, one needs to meticulously avoid using words without considering potential misinterpretations. Use of ambiguous terminology creates a risk of misunderstanding; and a good attorney knows how to frame questions to solicit your response.

During one lengthy cross examination about outdoor cannabis cultivation and yields, I testified that *Sinsemilla* gives a higher usable yield than seeded cannabis because a third to three-fourths of the dry flower mass of a seeded plant is seed and gets subtracted from the usable weight. "Would it surprise you to know that a federal DEA study[8] found that seeded marijuana produces more bud?" That was a surprise, since I know the document includes two charts showing the exact opposite: *Sinsemilla* (literally, *without seed* in Spanish) produces a greater usable yield. I asked to see the document and he pointed to a mention that fertilized plants gave a higher yield. The report meant fertilized with plant nutrients – he interpreted that as seed pollination. I quickly made the distinction, to the dismay of said DA.

Opinion versus speculation

An expert may be able to bolster a legal theory as long as the case materials are thorough, accurate, clear and well-organized (see Box 9.4). A court expert is allowed to give opinion but must be ready to support that claim. In other words, an opinion has to be an educated guess, not just some vague possibility. For example, one might opine that a quantity of marijuana was possessed for personal use, rather than sales, based on the amount, quality, packaging, and lack of sales indicia.

> **Box 9.4:** Representative legal issues frequently encountered in cannabis cases.
>
> **Typical legal issues:**
> - Hemp versus marijuana
> - Processing and quality
> - Ingestion and quantity
> - Indicia of personal use
> - Indicia of criminal activity, or lack thereof
> - Odor and detection
> - Copyrights and patents
> - Loss valuations

---

8 ElSolhy, M. et al., *Cannabis Yields*, NIDA/DEA report. National Institute on Drug Abuse. 1992.

You cannot speculate that something happened simply because "anything is possible." To dispute an "anything is possible" attack from opposing council, try to qualify what "possible" means. For example, one officer said he had precisely pinpointed an illegal marijuana grow site while driving down the street. I opined that the factors of legal gardens, hemp odor, distance, air currents, etc., all made his claim effectively impossible. "But anything is possible, right?" challenged the prosecutor. "Maybe, but not within these circumstances and variables."

Dueling experts

Often in criminal court, and virtually always in civil court, attorneys engage in a battle of dueling experts (see Box 9.5). Focusing your scope of testimony on direct examination reduces the issues the other side can raise on cross. However, they can bring in a wide variety of things as rebuttal or for impeachment, such as something attributed to you in an interview or in a study that you cited or published. Opposing counsel will try to give their expert more credibility. "Is it fair to say that someone at the scene would know better than someone who was not there?" Response: "It all depends on how well they understand what they are looking at."

Box 9.5: Typical reasons attorneys bring experts into legal proceedings.

There are a number of reasons why an attorney might bring in an expert:
- Draw upon their technical and legal experience to support the case
- Test the merits and weaknesses of various legal or factual theories
- Counter opposing experts
- Demonstrate case deficiencies
- Signal opposing counsel that they are in for a fight
- Create leverage for negotiations
- Mitigate jury and judicial bias

A judge may restrict the number of experts who testify, forcing counsel to pick and choose which are more important to their legal theory. It pays to have experience that sets you apart from others, like work in clinical studies or having interviewed people about their relevant experiences. I brought three specialties to the stand: (1) I grew and processed cannabis legally in Holland, not illegally in the United States, so I broke no federal laws. (2) Medical marijuana activism allowed me to speak with thousands of patients and consumers and scores of growers, caregivers, physicians, legislators, and attorneys, often at conferences, including voluntary or mandatory continuing medical education (CME)[9] and continuing legal education (CLE) trainings.[10] (3) I was uniquely informed in the minutiae of relevant state laws because I had trained on them, lectured on them, played a role in their passage, or been cited in major case law decisions.

Prepare to be attacked

Attorneys work hard to exploit any perceived weakness and make the other side look bad. You may find yourself repeatedly saying under oath, "No, I was not present at the incident. No, I am not a physician. No, I am not a policeman." It is essential to explore all potential weaknesses of the case in the early stages of your engagement. You don't want to be on the stand and face a question that you have not yet discussed with counsel. If it makes tactical sense, suggest they add another expert to cover specific issues of the case.

Before a deposition or testimony, you may be required to provide reciprocal discovery to back up your training and opinions. Likewise copies of previous depositions may be subpoenaed if you retain copies for your records. Counsel may pursue you with, "How many hours of training did you spend on this particular issue?" and ask to discover any class or training materials. If you mention a publication, counsel might pull out a copy of it, go to any page and start asking you questions. One attorney quoted an unrelated essay from a compendium that a medical expert had contributed to and suggested that it showed bias. It was not even that expert's opinion. The solution is, of course, to limit your cited sources, know your information well and testify about both academic and worldly experience.

Always be circumspect in your word choices and consider how terms can be misinterpreted. Recognize any weaknesses of your testimony and be prepared to explain. Think three questions ahead and always speak the truth, though it be uncomfortable. If opposing counsel asks about a written statement attributed to you, ask to read it in context. Ask to see the document in question. I've had DAs try to blindside me by reading aloud an out-of-context, one-line quote from a large court transcript and ask if those were my words. When I read the document, I found that the surrounding sen-

---

9 "Continuing medical education (CME) consists of educational activities that serve to maintain, develop, or increase the knowledge, skills, professional performance and relationships that a physician uses to provide services." They can be accredited by the AMA or groups like American Association of Continuing Medical Education. https://aacmet.org/cme/what-is-cme/.

10 Attorneys are required to take a certain number of hours of CLE course work per year in all but four states. In February 2017, the American Bar Association updated its *Model Rule for Minimum Continuing Legal Education and Comments* ("MCLE Model Rule"). https://www.americanbar.org/events-cle/mcle/modelrule/.

tences – omitted from of his quote – explained exactly why I said what I did, so I read that back to him before confirming the words were mine.

Counsel may ask a yes or no question that does not have a binary value, like the classic, "Have you stopped beating your wife?" Your answer should be, "I never did that;" but a judge might insist on "yes" or "no." You can try "neither" or decry the ambiguity of their question. Humor may help the situation. "The last time I beat my wife was in a game of checkers." A useful response is, "Yes/no, but with an explanation." Opposing counsel may not follow up but your retaining attorney can ask what you meant.

Conflicts and skeletons in the closet

When working in civil courts, the expert should keep a database to track all the plaintiffs and defendants to avoid a conflict of interest in their work. As a criminal defense expert, it's also generally beneficial to keep track of the attorneys, courthouses, judges, opposing counsel, and prosecutors you encounter, in case they show up at some other point in your career.

Always be front and center about anything that causes a conflict or could show you in a bad light. Sometimes it's good to handle a delicate issue, like your being paid, in direct than on cross examination and other times better to let it go but be ready, in case it comes up later. The question, "do you use cannabis?" is pretty likely to come up. The fact that I consume cannabis was a hot issue for the first 20 years of my career, since it was federally illegal and the thinking was that a judge, magistrate or jury would be prejudiced against me. Sworn to the truth, I embraced it and often got a few smiles by flipping my pony tail and saying, "Of course I smoke it, I'm a cannabis expert."

The "skeletons in my closet" have been my use of cannabis, public advocacy and defining (hence founding) *Cantheism*, the practice of sharing cannabis as a sacrament (Cannamaste).

Final notes

Keep detailed time logs when billing by the hour, especially for public defenders (PDs). Many private attorneys and clients prefer a flat rate for a report, a property or site inspection or a day in court, to contain costs. You may wish to work pro-bono on sympathetic cases, especially getting started, and I always consider the finances of each client.

Know who's responsible for paying your invoice and always get paid up front. It's not particularly fun to refund people's money – but it's a lot better than being stiffed. The more rushed a request, the riskier. When people need you, they want a lot of work from you but, once the matter settles, your invoice seems much less urgent. If you don't have the money in your pocket before then, you might be left hanging. Even if you win in small claims court, you still need to collect the money, so hope they come through but consider it *pro bono.*

PDs are extremely important in criminal law because they have budgets to pay expert witnesses and to file appeals to a higher court. A retained attorney is only retained for one phase of a case and is much less likely to appeal. Some of the most important appeals court rulings of my career were handled by PDs [16–19]. In federal and state officers, PDs have special requirements like needing official approvals, having budget caps, seeking discounts and payment after the case closes. Fortunately, this is one group you can rely on to pay you . . . as long as all the paperwork is filed correctly and on time. It's your due diligence to make sure that happens.

And due diligence is the name of the game for a cannabis expert witness.

References

[1]   Rule 702. Testimony by Expert Witnesses | Federal Rules of Evidence | US Law | LII / Legal Information Institute [Internet]. [cited 2025 Sept 3]. Available from: https://www.law.cornell.edu/rules/fre/rule_702

[2]   Oaksterdam University Courses [Internet]. [cited 2025 Sept 3]. Available from: https://oaksterdam.com/courses/

[3]   120 of the best college courses, degrees, and certifications for cannabis [Internet]. [cited 2025 Sept 3]. Available from: https://www.leafly.com/news/industry/best-cannabis-college-degrees-and-certifications

[4]   California Code, Evidence Code – EVID § 720 | FindLaw [Internet]. [cited 2025 Sept 3]. Available from: https://codes.findlaw.com/ca/evidence-code/evid-sect-720/

[5]   Daubert , V. Merrell Dow Pharmaceuticals, Inc., 509 U.S. 579 (1993). [Internet]. Available from: https://supreme.justia.com/cases/federal/us/509/579/case.pdf

[6]   Conrad, C. (2005). Cannabis Yields and Dosage. A Guide to the Production and Use of Medical Marijuana. Creative Xpressions, Creative Xpressions.

[7]   Brenneisen, R. (2007). Chemistry and Analysis of Phytocannabinoids and otherCannabisconstituents. In: ElSohly, M. (ed). Marijuana and the Cannabinoids. Humana Press, 17 49.

[8]   Russo, E.B., Marcu, J. (2017). Chapter Three Cannabis Pharmacology: The Usual Suspects and a Few Promising Leads. In: Adv Pharmacol. Vol. 80, 67–134.

[9]   Gilbert, A.N., DiVerdi, J.A. (2018). Consumer perceptions of strain differences in Cannabis aroma. PLoS ONE, 13(2), e0192247.

[10]   Armentano, P., Bacca, A., Vazquez, L. The Budtender's Guide: A Reference Manual for Cannabis Consumers and Dispensary Professionals. Internet Oaksterdam Press, Available from https://books.google.com/books?id=bMnGzwEACAAJ.

[11]   Cannabis Yields (1992). Drug Enforcement Agency [Internet]. Available from: https://www.scribd.com/document/61211679/1992-Cannabis-Yields-DEA

[12]   Rule 26. Duty to Disclose; General Provisions Governing Discovery | Federal Rules of Civil Procedure | US Law | LII / Legal Information Institute [Internet]. [cited 2025 Sept 4]. Available from: https://www.law.cornell.edu/rules/frcp/rule_26

[13]   Justice Manual | 165. Guidance for Prosecutors Regarding Criminal Discovery | United States Department of Justice [Internet]. [cited 2025 Sept 4]. Available from: https://www.justice.gov/ar chives/jm/criminal-resource-manual-165-guidance-prosecutors-regarding-criminal-discovery

[14]   Rule 16. Discovery and Inspection | Federal Rules of Criminal Procedure | US Law | LII / Legal Information Institute [Internet]. [cited 2025 Sept 4]. Available from: https://www.law.cornell.edu/ rules/frcrmp/rule_16

[15]   McPartland, J.M., Clarke, R.C., Watson, D.P. (2000). Hemp Diseases and Pests: Management and Biological Control. CABI, (CABI).

[16]   People v. Orlosky | 233 Cal. App. 4th 257 | Cal. Ct. App. | Judgment | Law | CaseMine [Internet]. [cited 2025 Sept 4]. Available from: https://www.casemine.com/judgement/us/ 5c55354e342cca0f8229f242

[17]   People v. Kelly – 47 Cal. 4th 1008, 222 P.3d 186, 103 Cal. Rptr. 3d 733 S164830 – Thu, 01/21/2010 | California Supreme Court Resources [Internet]. [cited 2025 Sept 4]. Available from: https://scocal. stanford.edu/opinion/people-v-kelly-33805

[18]   People v. Mower – 28 Cal.4th 457 S094490 – Thu, 07/18/2002 | California Supreme Court Resources [Internet]. [cited 2025 Sept 4]. Available from: https://scocal.stanford.edu/opinion/people-v-mower -32256

[19]   PEOPLE v. URZICEANU (2005). | FindLaw [Internet]. [cited 2025 Sept 4]. Available from: https://caselaw.findlaw.com/court/ca-court-of-appeal/1265236.html

Fredrick J. Goldstein

Cannabis as medicine: established uses and known risks

Introduction

Marijuana use has increased significantly in America over the past two decades, driven by the plant's therapeutic potential and facilitated by the majority of states legalizing its sale for both recreational and medical purposes. While recreational use typically requires only basic identification for purchase, medical cannabis programs often involve more rigorous processes including physician authorization. This widespread state-level acceptance, however, exists in direct conflict with federal law, where marijuana remains classified as a Schedule I substance under the Controlled Substances Act – a designation reserved for drugs deemed to have "no currently accepted medical use in the United States, a lack of accepted safety for use under medical supervision, and a high potential for abuse" [1]. This fundamental contradiction between federal policy and emerging evidence becomes even more striking when viewed against the backdrop of cannabis's extensive medicinal history and the growing scientific understanding of its complex pharmacology.

Cannabis sativa in the form of seeds and other parts of the plant has been used in China for pain and mental conditions for almost 2,000 years [2]. The initial isolation of delta-9-tetrahydrocannabinol (THC) from marijuana was reported more than 60 years ago [3]. However, the chemical profile of cannabis strains is complex; more than 550 compounds have been identified including >140 cannabinoids, 120 terpenes, hydrocarbons, sugars, ketones, aldehydes, fatty acids, amino acids, lactones, and vitamins, many of which probably contribute to the differential pharmacological effects per strain [4].

An excellent review [5] presented material on cannabinoid receptors, now classified as CB1 (mostly in the central nervous system [CNS]) and CB2 (mostly in the peripheral nervous system), and the extensive reach of the endocannabinoid system (ECS). This system is involved in at least 31 physiologic functions spanning multiple organ systems including the brain, cardiovascular (CV) system, gastrointestinal tract, liver, reproductive system, and skeletal muscle (see Table 1). Information on many of the major THC-induced therapeutic actions and adverse reactions related to these physiologic functions is presented herein.

© 2026 Walter de Gruyter GmbH, Berlin | https://doi.org/10.1515/9783111476162-012

Table 1: Physiologic functions of the endocannabinoid system.

Brain
      Learning, memory, cognition
      Anxiety and depression
      Appetite and food intake
Cardiovascular system
      Vasodilation
      Cardiac function
Emesis
      Nausea
GI tract
      Motility
      Intestinal barrier function
Liver
      Insulin resistance
Nociception
Reproductive system
      Fertility
Skeletal muscle

Therapeutic effects

Therapeutic applications of cannabis and its constituents have been extensively studied across multiple medical conditions, with varying degrees of clinical evidence supporting their efficacy. This section examines major therapeutic categories in which cannabinoids have demonstrated clinical utility, including pain management, nausea and vomiting control, spasticity reduction in neurological disorders, appetite stimulation, and seizure management. While the quality and scope of evidence varies among these applications, each represents an area where cannabinoids have shown sufficient promise to warrant clinical investigations.

Analgesia

Several decades ago, it was proposed that motor, cognitive, analgesic, and autonomic effects are mediated by CNS cannabinoid receptors; these sites are found in brain areas which regulate pain, stress, and emotion [6]. Subsequently, it was demonstrated that endogenous cannabinoid system ligands and exogenous cannabinoid receptor ligands are involved in modulation of pain, i.e., CB1 receptors on central neurons and CB2 sites on immune system cells [7, 8].

In patients with chronic pain from diabetic neuropathy, a 6-week crossover trial on 40 patients was conducted to evaluate analgesia from nabilone (similar molecule to THC; Cesamet®). Compared to a placebo, this drug produced a significant decrease in pain scores (11-point Visual Analog Scale [VAS]) and improved quality of sleep. Drowsiness and fatigue were the most frequent adverse effects [9].

A clinical study evaluated the effectiveness of vaporized cannabis as an analgesic in 39 patients with peripheral neuropathic pain. They followed a standardized procedure for inhaling medium dose (3.53% THC), low dose (1.29% THC), or placebo. A standard 11-point visual analog scale was employed to evaluate the degree of pain relief, a reduction of 30% from baseline being an endpoint. Results demonstrated that analgesia was produced, the low dose being equal to the medium dose and both were more effective than placebo. There were psychoactive effects, e.g., feeling high, stoned, drunk, and/or impaired; however, they were minimal, well-tolerated, and readily reversible within 1–2 h. However, a limitation of this research is that the actual doses of THC were not given [10].

Employing the GRADE (Grading of Recommendations Assessment, Development and Evaluation) system [11], a review rated the overall quality of research evidence for risk of bias, publication bias, and imprecision in reports on medical uses of cannabinoids. After examination of 79 trials (6,462 participants), moderate-quality evidence found that cannabinoids could be beneficial for the reduction of chronic neuropathic or cancer pain. However, associations did not reach statistical significance in all studies. Adverse reactions documented included dizziness, dry mouth, nausea, fatigue, somnolence, euphoria, vomiting, disorientation, drowsiness, confusion, loss of balance, and hallucinations [12].

In a 4-week, randomized, placebo-controlled study on neuropathic pain, a topically delivered cannabidiol (CBD) oil was evaluated. There were 29 patients, 14 randomized to placebo, and 15 to treatment which contained 250 mg CBD. The placebo group was permitted to crossover into the treatment group after 4 weeks. As a result, those using transdermal CBD experienced a statistically significant decrease in intense pain, sharp pain, and cold and itchy sensations compared to placebo. There were no reported adverse events [13].

Administering dronabinol (FDA-approved synthetic form of THC) for chronic neuropathic pain in multiple sclerosis (MS) patients showed that this drug, as long-term analgesic treatment, was both an effective and safe pain reliever, with adverse events decreasing over time. Study participants did not exhibit signs of THC adverse effects within a maximum employed dose of 15.9 mg [14].

Results from a pilot study have shown that dronabinol has efficacy in reducing neuropathic pain in several conditions, e.g., MS, peripheral diabetic neuropathy and temporomandibular joint disorder. In this study, the seven patients attended weekly visits for vital-sign monitoring while also completing a 0–10 pain diary three times daily and documenting their daily pain medication use. Weeks 1 and 2 were baseline; weeks 3–6 were the up-titration process. Oral dosage started at 5 mg at bedtime, increasing weekly to the maximum of 20 mg in week 6. The subjects experienced an average pain reduction of 44.9%, and most used less gabapentin, carbamazepine, and ibuprofen [15].

Although several pain studies employed specified doses of THC, CBD, and nabilone, there is a lack of such specificity in clinical research on medical marijuana. A recent review of 440 patients who had qualified for a state- and physician-approved medical marijuana card showed that at 6 months, 38.6% had increases in analgesia, function, or global impression of change. There also was a decrease in morphine milligram equivalents of 39.3%. However, an important concluding statement was a limitation in that actual doses of THC and CBD were unknown [16].

Antiemetic

Although it is generally known that marijuana in clinical situations can reduce nausea and vomiting from antineoplastic therapy, there are few controlled studies to show this effect.

An early placebo-controlled, randomized, double-blind study evaluated antiemetic efficacy of THC in patients receiving many different antineoplastics that cause nausea and vomiting, including adriamycin, procarbazine, and high-dose either cyclophosphamide or methotrexate. Patients served as their own controls, i.e., a crossover procedure. Oral doses of THC, calculated as 10 mg of THC per square meter of body surface area, were either 15 mg (19 patients) or 20 mg (3 patients) given three times over a 10-h span. Of the 20 patients who completed the study (e.g., one died during the study period), 14 showed either a complete or partial (minimum of 50% reduction compared to placebo) antiemetic effect. A "high" was experienced by 81% of those during their THC dosing; reactions included easy laughing, elation, heightened awareness, and mild aberrations of fine-motor coordination. Somnolence also occurred. A patient who received 60 mg of THC experienced toxic psychosis which included paranoia, panic and frightening visual hallucinations [17].

Using vomiting induced by syrup of ipecac, an investigation was conducted in normal subjects using smoked marijuana to affect this pharmacologic emesis. Marijuana cigarettes contained either 8.4 or 16.9 mg of delta-9-THC; the third group was given 8 mg of ondansetron. Compared to placebo, "queasiness" was significantly reduced and vomiting was slightly decreased by marijuana. In contrast, ondansetron completely eliminated the ipecac-induced emesis [18].

A THC:CBD combined product was evaluated as antiemetic for refractory chemotherapeutic-induced nausea and vomiting (CINV). This was a multicenter, randomized, double-blind, placebo-controlled, phase II/III trial designed to determine the efficacy of an oral THC 2.5 mg/2.5 mg CBD combination added to the current anti-emetic therapy to prevent refractory CINV. The percent of the 81 patients who had a complete response during the entire 120 h of chemotherapy was the primary endpoint.

Results showed that a complete response was improved from 14% up to 25%; similar effects were noted on absence of emesis, use of rescue medications, absence of significant nausea, and summary scores for the Functional Living Index-Emesis (FLIE). However, 31% developed moderate or severe cannabinoid-related adverse events such as sedation, dizziness, and disorientation. The conclusion was that adding a THC:CBD combination to routine antiemetics produced less nausea and vomiting but additional side effects [19].

A recent study measured how a wide range of cannabis products affected nausea intensity. Employing the Releaf App, 886 people completed 2,220 cannabis anonymously recorded, self-administration sessions for the treatment of nausea during a 3-year period. They recorded the characteristics of self-administered cannabis products and baseline symptom intensity levels before tracking real-time changes in the intensity of their nausea. A major result was that within an hour of using their respective marijuana products, 96.4% experienced symptom relief; on a 0 to 10 VAS, with the average reduction in symptom intensity being 3.85 points. As indicated by the authors, a limitation was not capturing conditions creating the nausea and vomiting [20].

A recent excellent review evaluated 26 clinical studies using search terms of cannabinoids (cannabis, dronabinol, cannabinoids, cannabidiol, nabilone, and THC) and CINV. Meta-analysis was performed for each endpoint and composite ones which amalgamated existing endpoints. Compared to placebo, cannabinoids were shown to have superior overall reduction of CINV but no differences were detected between cannabinoid and active treatment alternatives (primarily prochlorperazine, haloperidol, and chlorpromazine). A conclusion was that limited evidence exists for the efficacy of cannabinoids for CINV in current employment of triple and quadruple antiemetics [21].

While Sallan and colleagues [17] demonstrated that oral THC significantly reduced CINV in cancer patients – with antiemetic effects observed in 14 of 20 THC courses compared to none of 22 placebo courses – a subsequent study from the US Surgery Branch of the National Cancer Institute yielded contrasting results. Chang and colleagues [22] enrolled eight patients with resected soft-tissue sarcomas in a randomized, double-blind, placebo-controlled trial to evaluate oral THC's effectiveness against nausea and vomiting caused by adriamycin or cyclophosphamide therapy, with patients serving as their own controls and receiving doses based on 10 mg/m^2 of body surface area. In this study, oral THC proved ineffective. However, patients who continued to vomit were subsequently given marijuana cigarettes containing 17.4 mg of THC each – a different route of administration that may have contributed to the divergent outcomes between these two studies

Since results from these all of these investigations are somewhat inconclusive, additional studies will be required to establish a role, if any, for marijuana as an antiemetic agent.

Antispastic: MS

A study was conducted on MS patients who were resistant to antispastic treatment. Data were collected from 30 Italian centers specializing in such treatment, focusing on THC:CBD oromucosal spray (Sativex®; nabiximols; not available in the United States) efficacy and tolerability in those with drug-resistant spasticity. Eligibility criteria for recruitment were: ≥18 years of age, minimum of moderate spasticity (numerical rating 0–10 scale [NRS] score ≥4) who had not responded to common antispastic drugs (none were listed). Evaluation of patients was at baseline and following 4 weeks of treatment. Data were also collected about any meaningful improvements in six primary spasticity-related symptoms. There were 1,615 patients enrolled with 1,432 completing a 1-month trial period. The average use was 6 puffs daily; each spray consisted of 2.7 mg THC and 2.5 mg CBD. In this group, 70.5% (1,010 patients) reported a reduction in NRS score of ≥20% compared with baseline. It was also observed that 43.8% (627patients) experienced an improvement in at least one spasticity-related symptom, i.e., cramps/nocturnal spasms, bladder control, pain, sleep quality, clonic movements, and mood [23].

A review of cannabinoids in MS [24] highlights both preclinical and clinical findings, showing that CBD alone and in combination with THC demonstrates anti-inflammatory and symptomatic benefits, though clinical outcomes are mixed. The pertinent findings are below:
- CBD by reducing spinal cord T-cell infiltrates has anti-inflammatory actions which, in animal models of MS can retard disease progression.
- CBD in combination with THC has greater efficacy in such animal models.
- Clinical studies of 1:1 THC:CBD are less positive in MS patients but have an efficacy close to current front-line treatments, e.g., baclofen [GABA-B receptor agonist] and tizanidine (α-2-adrenergic receptor agonist).
- Cannabinoids may be of benefit in treating neuropathic pain in MS when added to the existing analgesic therapy but may result in substantial cognitive dysfunction.

Appetite stimulant

It is well-known that recreational marijuana, smoked or oral, increases appetite and consumption of food, mostly high-calorie products, e.g., brownies, "munchies" is the common non-official term. However, formal clinical studies have shown mixed results.

A 6-week prospective investigation was designed to compare efficacies of cannabis extract (CE), THC, and placebo on appetite and quality of life (QOL) in 164 patients with cancer-related anorexia-cachexia syndrome. Increases in appetite were reported in all three groups: CE – 73%; THC – 58%; placebo – 69%. Due to lack of differentiation, an independent data review board recommended termination [25].

In a study on 469 advanced cancer patients, dronabinol was compared to megestrol acetate for treatment of anorexia. Doses studied were (1) oral megestrol acetate 800 mg/day liquid suspension plus placebo, (2) oral dronabinol 2.5 mg twice a day plus placebo, or (3) both agents. Eligibility included loss of appetite and decrease in weight of five or more pounds in the previous 2 months. More patients reported improvement in appetite and weight gain with megestrol compared to dronabinol, i.e., 75% versus 49% ($p = 0.0001$) [26].

However, a large review study reported the efficacy of cannabinoids. It was focused on five randomized controlled trials (RCTs) in peer-reviewed journals which compared medicinal cannabis with placebo and another intervention. Compared were dronabinol, nabilone, and CE with placebo ($n = 4$) or megestrol acetate ($n = 1$). Only one investigation produced positive results: dronabinol significantly improved chemosensory perception, taste of food, premeal appetite, and proportion of calories consumed as protein compared with placebo [27].

The mechanism by which this effect occurs has been studied. An investigation was undertaken to elucidate the effects of different routes of cannabis administration on peripheral concentrations of appetitive and metabolic hormones in 20 users. It was a randomized, crossover, double-blind, placebo-controlled study. Oral, smoked, and vaporized cannabis, or placebo was given. The calculated dose of THC from each route was approximately 50.6 mg. Blood samples of endocrine markers were measured: total ghrelin, acyl-ghrelin, leptin, glucagon-like peptide-1 (GLP-1), and insulin. Among mechanistic results in the THC groups compared to control were (1) lower insulin levels, suggesting less glucose entering cells thus increasing the need to consume more calories, and (2) lower GLP-1 levels; this substance retards gastric emptying and decreases appetite via peripheral (e.g., stomach) and central (e.g., hypothalamus) mechanisms, an effect which would also contribute to increased appetite [28].

Anticonvulsant

Although THC is effective in as an antiemetic and analgesic, many studies on cannabis-related efficacy in treating epilepsy are also linked to CBD. A recent excellent review [29] of cannabinoid use in epilepsy describes how THC primarily exerts anticonvulsant effects through CB1 receptor activity, while CBD acts via alternative pathways. Evidence for CBD is strongest in treatment-resistant epilepsies such as Dravet and Lennox-Gastaut syndrome, with further details and clinical considerations outlined below:
- THC acts through the ECS especially on CB1 receptors to produce its anticonvulsant activity.
- CBD does not bind to CB1 or -2; its anticonvulsant activity is most likely generated by reducing both oxidative stress and neuroinflammation while increasing GABA activity.

- CBD is effective in the anticonvulsant-resistant Dravet and Lennox-Gastaut syndromes which are severe and rare types of epilepsy. Epidiolex®, CBD, was approved by the FDA as the initial cannabis product for treatment of these serious conditions. It is well-tolerated, although the most frequent adverse reactions include increased gastrointestinal motility, somnolence and reduced appetite. Hepatotoxicity is not common but liver enzymes should be monitored during therapy.
- CBD may also serve as a prophylactic agent in epilepsy based upon these pharmacologic actions.

While the therapeutic applications of cannabis demonstrate significant clinical potential across multiple medical conditions, the use of cannabinoids is not without considerable risks. The same pharmacological mechanisms that produce beneficial effects can also generate substantial adverse reactions. A comprehensive understanding of these adverse effects is essential for clinicians considering cannabinoid therapy and for patients making informed decisions about cannabis use.

Adverse effects

Adverse effects of cannabis use span multiple organ systems and can range from mild, transient reactions to serious, potentially life-threatening complications. The following sections examine the major categories of adverse effects associated with cannabinoid use, including CV toxicity, neurological and psychiatric complications, gastrointestinal disorders, reproductive harm, and oncological risks. Understanding these adverse effects is crucial for risk-benefit assessment in both medical and recreational cannabis use.

Cardiovascular

Blood pressure

A United States National Health and Nutrition Examination Survey (NHANES) spanning 8 years (2005–2012) employed a computer-assisted self-interview process to assess cannabis use and blood pressure in a study population consisting of non-institutionalized community residents. The demographic data indicated that in comparison with never-users, recently active cannabis users tended to be younger, and non-Hispanic white males, with an income below the federal poverty level. From an analysis of 12,426 participants, a modest association between recent cannabis use and increased systolic blood pressure was observed; there was no such connection to diastolic blood pressure. A limitation is that there were no data on the frequency of cannabis use beyond 30 days before the interview, so it was not possible to determine if the relationship between blood pressure and cannabis use was short term or long term in nature [30].

In contrast, results from the 2009–2018 NHANES indicated that chronic cannabis used monthly for more than 12 months does not produce a clinically significant increase in systolic blood pressure. This was an observational study that analyzed cross-sectional data from the NHANES for five 2-year cycles; data were from 9,783 middle-aged adults from 35 to 59 years old. Strong points include a relatively large nationally-representative sample size; it was balanced with sociodemographic data and employed a standardized objective outcome measure [31].

While controversial, most studies show a reduction in blood pressure. As a representative investigation, a study in the United Kingdom surveyed 91,161 volunteers; cannabis use status was determined by a questionnaire requesting self-selection of being a heavy, moderate, or low usage, or a never-user. Results were that among lifetime heavy cannabis there was a reduction in systolic blood pressure, diastolic blood pressure, and heart rate, with a greater effect among women [32].

Coronary heart disease (CHD), myocardial infarction (MI), and stroke

A retrospective study of 2,097 patients presenting with a MI for ≤50 years was conducted at two academic hospitals from 2000 to 2016. Through a review of records for patient-reported or lab-screened substance misuse during the week before MI, a determination of substance abuse was made. Vital status was indicated by the Death Master File of the Social Security Administration. After reviewing electronic health records and death certificates, the cause of death was adjudged. Higher CV and all-cause mortalities were associated with the use of cocaine and/or marijuana [33].

An investigation combined 2016–2020 Behavioral Risk Factor Surveillance System (BRFSS) data from two US territories and 27 states that participated in the cannabis module during at least one of these years. The BRFSS process involves a telephone survey of a representative sample of US adults that collects information on risk factors, chronic conditions, and healthcare access. It was controlled for tobacco smoking status, age, sex, race and ethnicity, body mass index, diabetes, alcohol use, educational attainment, and physical activity. Data were gathered from 434,104 participants from 18 to 74 years old who responded to this question, "During the past 30 days, on how many days did you use marijuana or hashish?" Information collected for this time period days indicated that use of cannabis use was associated with MI, stroke, and composite outcomes of CHD, MI, and stroke. A dose-response was observed with higher risk occurring with more days of use [34].

A retrospective cohort study was conducted on a total of 4,636,628 adults who were relatively healthy and ≤50 years of age; 2.01% (93,267) were cannabis users with 97.99% (4,543,361) being non-users. Data were obtained from the TriNetX health research network, an electronic medical record system with de-identified patients. In this investigation, information was obtained from 53 US healthcare organizations. Both groups had no significant baseline comorbidities including, but not limited to, tobacco use, hypertension, hyperlipidemia, coronary artery disease (e.g., prior MI or history of coronary interventions), and diabetes mellitus or hemoglobin A1C > 7. Evidence provides a link of cannabis with CV toxicity manifested as MI, ischemic stroke, HF, and mortality. Conclusions are that cannabis is a significant, independent, novel, and unrecognized risk for such physiological damage [35].

Cannabis arteritis

An etiologic factor in such cardiotoxicity may be linked to cannabis-induced vascular damage classified as cannabis arteritis. In a case report, a 36-year-old man who was a cannabis user presented with digital necrosis on his right foot. There were no abnormal laboratory findings. An arteriogram showed distal segmental lesions and popliteal arterial occlusion. Worsening symptoms were associated with continued development of this condition in his left leg which ultimately required amputation. Although this was only one user, the authors concluded from the arteriogram that such pathology was also relatively similar to thromboangiitis obliterans in its clinical aspects. It was concluded that cannabis arteritis may be an under-diagnosed cause of juvenile peripheral obstructive arterial disease [36]. There are at least three additional case reports of this adverse phenomenon [37–39], although there is no review or case series available, this adverse reaction observed in four chronic marijuana users although rare identifies a serious toxic effect.

Nervous system

Central

Usual pharmacologic positive effects sought from acute marijuana include pleasant feelings, relaxation, laughter, and mild euphoria; hunger (a/k/a "munchies") while not necessarily a goal, is welcomed. Undesired reactions include reductions in psychomotor function. Since users recognize their impairment, they tend to drive slower as a compensatory process. However, they are then arrested by law enforcement officers because their car speed is substantially lower than required, and on a high-speed expressway, very noticeable and dangerous [40, 41].

However, there are negative CNS consequences of marijuana use. An early article documented various marijuana-induced central effects which remain valid today [42]. Three categories were presented primarily based upon the physician-author's direct clinical experience:

Persons without history of mental disorder; have never taken hallucinogens
Mild effects such as nausea, headache, and transient paranoia may occur but most who use it do not experience any subjective effects. However, there are several reactions that may ensue:

Panic reactions
Some users – including new ones – who have some concern regarding initiation the of this substance experience the physical and mental effects; they believe that they are near death or losing their minds. However, they calm down when reassured by those who are regular users (or doctors, if they are seen in the emergency department) that such actions are due to the drug, and will end as the drug clears from the body.

Simple depression
This effect can be also be precipitated by marijuana; it is more likely in new users and those with obsessive-compulsive disorder. This adverse effect is similar to transient neurotic depression, which is not related to this drug, and ends spontaneously.

Toxic psychosis
Any person who consumes a very large dose may develop psychosis. The clinical presentation is similar to a high fever delirium. Such reactions are also self-limiting.

Persons without a history of mental disorder; have previously taken hallucinogens

Reoccurrence of hallucinogenic experience
Persons who have also used psychedelic drugs may experience the same type of reactions when they use only marijuana, i.e., a "flashback." In this situation, these effects are benign and will disappear. However, flashbacks can also appear independently of marijuana use.

Precipitation of delayed psychotic reactions to hallucinogens
On occasion there is a psychotic reaction months after a last use of marijuana.

Persons with a history of psychosis
In some, a psychotic episode develops when usual, i.e., not toxic, amounts of marijuana are consumed. A possibility is that these persons have a low threshold for this serious adverse reaction.

Behavioral, cognitive, and endocrine effects were evaluated in 22 healthy individuals following intravenous doses of 2.5 mg and 5 mg of THC in a 3-day, double-blind, randomized, counterbalanced study. Results included: (1) production of schizophrenia-like positive and negative symptoms; (2) alteration in perception; (3) euphoria; (4) anxiety; (5) disruptions in immediate and delayed word recall; (6) impairment in performance on tests of distractibility, verbal fluency, and working memory; and (7) elevated plasma cortisol. There was no impairment in orientation. From these data, it was concluded that THC can create a wide range of transient symptoms, behaviors, and cognitive deficits in healthy individuals that resemble certain features of endogenous psychoses [43].

While external THC administration can induce psychotic symptoms in healthy individuals, emerging research suggests that people with psychotic disorders may already have naturally elevated endocannabinoid activity, pointing to an intrinsic biological connection between the ECS and psychosis. In an excellent review by Minichino [44] of 18 studies designed to analyze blood and cerebrospinal fluid (CSF) measures of the ECS in psychoses, the results showed:
- significantly higher concentrations of anandamide in the CSF of patients than controls;
- significantly higher anandamide levels in the blood of patients compared to controls;
- significantly higher expression of CB1 cannabinoid receptors on peripheral immune cells in patients compared with controls.

A conclusion was that it appears possible to use clinically relevant ECS biomarkers in the CSF and blood to identify psychotic disorders.

A recent evaluation of 32 systematic reviews and meta-analyses on the connection between cannabis use and psychosis-related outcomes across the psychosis continuum concluded that the use of marijuana can create subclinical psychosis states (psychotic-like experiences) and traits (schizotypal personality) in the healthy population, as well as with an earlier onset and development of psychosis [45].

Unfortunately, antipsychotic therapy fails in the presence of marijuana use; a contributing factor is known medication nonadherence in psychosis. Within the pharmacologic realm, an extensive review presented data from many studies indicating that such unsuccessful treatment is linked to non-dopaminergic pathophysiology as occurs in patients with a treatment-resistant illness. Antipsychotics, as dopamine antagonists, act to counter increased presynaptic dopamine function in psychosis. However, regular use of marijuana may downregulate the dopaminergic system while increasing central glutamine. This is similar to treatment-resistant patients where poor efficacy to antipsychotic therapy is associated with elevated glutamate levels. In support, there is evidence that one dose of THC increases levels of striatal glutamate in controls [46].

Psychomotor

Psychomotor function can be reduced by cannabis. An investigation was conducted on 191 persons; demographics were 61.8% male with a mean age of 29.9 years and a mean use of 16.7 days in the previous 30 days. This double-blind, placebo-controlled, parallel, randomized laboratory clinical trial employed driving simulator variables. Treatments were 5.9% or 13.4% THC cannabis smoked ad libitum or placebo, with a primary end point of a Composite Drive Score. The result was that smoking cannabis ad libitum by regular users caused decreases in simulated driving [47].

An extensive review of THC-related motor vehicle accidents involved several published meta-analyses which included evaluation of 28 estimates from 21 observational studies. Lab results were employed to confirm the presence of THC. It was concluded that approximately 20–30% of traffic crashes occurred due to cannabis use [48].

Gastrointestinal

Hyperemesis

Although THC can reduce nausea and vomiting following acute therapeutic administration, chronic use of marijuana, especially with high doses seen in recreational use, can produce the cannabis hyperemesis syndrome (CHS). It has been defined as an aspect of a "functional gut-brain axis disorder characterized by bouts of episodic nausea and vomiting worsened by cannabis intake" [49], and was initially characterized clinically in 2004 [50].

A case series report on 98 patients at the Mayo Clinic proposed the following clinical criteria to characterize the CHS:
- severe cyclic nausea and vomiting;
- resolution of symptoms with cannabis cessation;
- symptom relief with hot showers or baths;
- abdominal pain (epigastric or periumbilical);
- at least weekly marijuana use.

Additional features to support this diagnosis include age younger than 50, loss of weight more than 11 pounds, major occurrence, and normal bowel movements with negative endoscopic, radiographic, and lab results [51, 52].

Acute pharmacologic management is less than ideal with varied results from serotonergic agents, e.g., ondansetron, and dopamine antagonists, e.g., haloperidol and metoclopramide. Fluid restoration is important. A non-drug process followed by many users is to take multiple hot showers, which effectively ameliorates the severity of CHS [49]. A possible explanation for the efficacy of such an elevated temperature process may be chronic stimulation of vascular CB1 receptors. These sites in the splanchnic circulation cause vasodilation. Short-term relief of symptoms could be associated with redistribution of blood flow from the gut to the skin [53].

Teratogenicity

A population-based retrospective cohort study focused on 316,722 pregnancies with a duration of 20 weeks or longer; subjects were screened for prenatal cannabis use. It was defined here as any self-reported administration during the initial stage of pregnancy or a positive lab result.

The frequency of cannabis use was 0.6% daily [1,930], 0.7% weekly [2,345], 0.5% monthly or less [4,892], and 3.4% unknown [10,886]. Results indicated prenatal cannabis was associated with the following adverse maternal health outcomes during pregnancy: gestational hypertension, preeclampsia, and placental abruption [54].

A retrospective study was conducted to examine whether cannabis use during pregnancy is linked to adverse outcomes in their children, using data collected from the baseline session of the Adolescent Brain and Cognitive Development Study. Evaluated were 11,875 children from 22 sites across the United States, aged from 9 to 11 years old, in addition to a parent or caregiver. Data showed in these children increased psychopathology characteristics (i.e., psychotic-like experiences, attention, thought, and social problems), sleep disturbances, body mass index changes, as well as changes in gray matter volume and reduced cognition. This indicates that prenatal cannabis is associated with greater risk for psychopathology during middle childhood, and that use of this substance should be avoided during pregnancy [55].

Cancer: head and neck

While the oncological risks of cannabis have been debated, emerging evidence suggests potential associations with head and neck malignancies that warrant clinical attention. In a large multicenter cohort study, clinical records over a 20-year span were reviewed for US adults with no prior history of head and neck cancer (HNC) who presented with and without cannabis-related disorder in outpatient hospital visits. The cannabis group was composed of 116,076 persons, of which 44.5% were female with a mean age of 46.48 years. The non-cannabis-related disorder cohort had 3,985,286 persons that was 54.5% female with a mean age of 60.8 years. Despite the cannabis group being significantly younger, in all sites, the rate of new HNC diagnoses was higher in the cannabis-related disorder group. Specifically, increased rates were in oral, oropharyngeal, and laryngeal cancers. While this observational study cannot establish causation and may be influenced by confounders, these findings suggest that chronic cannabis exposure may represent an under recognized risk factor in this type of cancer [56].

Conclusion

This review has provided information regarding the clinical benefits from marijuana use in addition to mild and serious toxic effects from such exposure. In this regard, it is of significance to indicate that in May 2024, the Department of Justice (DOJ) proposed in the Federal Register that marijuana be reclassified in the Controlled Substances Act (CSA) from Schedule I to Schedule III [57]. The DOJ decision was in agreement with the United States Department of Health and Human Services position that marijuana has (1) currently accepted medical uses and (2) a lower potential for abuse and lower levels of physical or psychological dependence than other controlled substances. Most recently, in April 2026, the DOJ moved forward with rescheduling medical marijuana from Schedule I to Schedule III.

References

[1]   Drug Enforcement Administration, Diversion Control Division. (n.d.). Controlled Substance Schedules. U.S. Department of Justice. Retrieved September 3, 2025, from. https://www.deadiver sion.usdoj.gov/schedules/schedules.html.

[2]   Brand, E.J., Zhao, Z. (2017). Cannabis in Chinese medicine: Are some traditional indications referenced in ancient literature related to cannabinoids. Frontiers in Pharmacology, 8(108), https://doi.org/10.3389/fphar.2017.00108.

[3]   Gaoni, Y., Mechoulam, R. (1964). Isolation, structure, and partial synthesis of an active constituent of hashish. Journal of the American Chemical Society, 86, 1646–1647.

[4]   Wishart, D.S., Hiebert-Giesbrecht, M., Inchehborouni, G., Cao, X., Guo, A.C., LeVatte, M.A., Torres-Calzada, C., Gautam, V., Johnson, M., Liigand, J., Wang, F., Zahraei, S., Bhumireddy, S., Wang, Y., Zheng, J., Mandal, R., Dyck, J.R.B. (2024). Chemical composition of commercial cannabis. Journal of Agricultural and Food Chemistry. 72(25), 14099–14113. https://doi.org/10.1021/acs.jafc.3c06616.

[5]   Zou, S., Kumar, U. (2018). Cannabinoid receptors and the endocannabinoid system: signaling and function in the central nervous system. International Journal of Molecular Sciences, 19(3), 833. https://doi.org/10.3390/ijms19030833.

[6]   Herkenham, M., Lynn, A.B., Johnson, M.R., Melvin, L.S., De Costa, B.R., Rice, K.C. (1991). Characterization and localization of cannabinoid receptors in rat brain: A quantitative in vitro autoradiographic study. Journal of Neuroscience, 11, 563–583.

[7]   Woodhams, S.G., Chapman, V., Finn, D.P., Hohmann, A.G., Neugebauer, V. (2017). The cannabinoid system and pain. Neuropharmacology, 124, 105–120. https://doi.org/10.1016/j.neuropharm.2017.06.015.

[8]   Chen, C.X., Cowan, A., Inan, S., Geller, E.B., Meissler, J.J., Rawl, S.M., Tallarida, R.J., Tallarida, C.S., Watson, M.N., Adler, M.W., Toby, K. (2019). Eisenstein Opioid-sparing effects of cannabinoids on morphine analgesia: Participation of CB1 and CB2 receptors. British Journal of Pharmacology, 176, 3378–3389. https://doi.org/10.1111/bph.14769.

[9]   Toth, C., Mawani, S., Brady, S., Chan, C., Liu, C., Mehina, E., Garven, A., Bestard, J., Korngut, L. (2012). An enriched-enrolment, randomized withdrawal, flexible-dose, double-blind, placebo-controlled, parallel assignment efficacy study of nabilone as adjuvant in the treatment of diabetic peripheral neuropathic pain. Pain, 153(10), 2073–2082. https://doi.org/10.1016/j.pain.2012.06.024.

[10] Wilsey, B., Marcotte, T., Deutsch, R., Gouaux, B., Sakai, S., Donaghe, H. (2013). Low-dose vaporized cannabis significantly improves neuropathic pain. The Journal of Pain, 14(2), 136–148. https://doi.org/10.1016/j.jpain.2012.10.009.

[11] Guyatt, G.H., Oxman, A.D., Vist, G.E., Kunz, R., Falck-Ytter, Y., Alonso-Coello, P., Schünemann, H.J., GRADE Working Group. (2008). GRADE: An emerging consensus on rating quality of evidence and strength of recommendations, BMJ (Clinical research ed., 336(7650), 924–926. https://doi.org/10.1136/bmj.39489.470347.AD.

[12] Whiting, P.F., Wolff, R.F., Deshpande, S., Di Nisio, M., Duffy, S., Hernandez, A.V., Keurentjes, J.C., Lang, S., Misso, K., Ryder, S., Schmidlkofer, S., Westwood, M., Kleijnen, J. (2015). Cannabinoids for medical use: A systematic review and meta-analysis. JAMA, 313(24), 2456–2473. https://doi.org/10.1001/jama.2015.6358.

[13] Xu, D.H., Cullen, B.D., Tang, M., Fang, Y. (2020). The effectiveness of topical cannabidiol oil in symptomatic relief of peripheral neuropathy of the lower extremities. Current Pharmaceutical Biotechnology, 21(5), 390–402. https://doi.org/10.2174/1389201020666191202111534.

[14] Schimrigk, S., Marziniak, M., Neubauer, C., Kugler, E.M., Werner, G., Abramov-Sommariva, D. (2017). Dronabinol is a safe long-term treatment option for neuropathic pain patients. European Neurology, 78(5-6), 320–329. https://doi.org/10.1159/000481089.

[15] Goldstein, F.J., Galluzzi, K.E., Brown, M., Smith, J., Lubeck, J. (2021). Analgesic effect of delta-9-Tetrahydrocannabinol on patients with chronic neuropathic pain during a four-week period: A pilot study. Research Week 2021, Philadelphia College of Osteopathic Medicine, DigitalCommons@PCOM, digitalcommons.pcom.edu/research_day/research_week_2021/research2021/94/

[16] Wasan, A.D., O'Connell, B., DeSensi, R., Bernstein, C., Pickle, E., Zemaitis, M., . . . Douaihy, A. (2022). The comparative effectiveness of medicinal cannabis for chronic pain versus prescription medication treatment. Pain, 10–1097. http://dx.doi.org/10.1097/j.pain.0000000000003506.

[17] Sallan, S.E., Zinberg, N.E., Frei, E. 3rd. (1975). Antiemetic effect of delta-9-tetrahydrocannabinol in patients receiving cancer chemotherapy. The New England Journal of Medicine, 293(16), 795–797. https://doi.org/10.1056/NEJM197510162931603.

[18] Söderpalm, A.H., Schuster, A., de Wit, H. (2001). Antiemetic efficacy of smoked marijuana: Subjective and behavioral effects on nausea induced by syrup of ipecac. Pharmacology, Biochemistry, and Behavior, 69(3-4), 343–350. https://doi.org/10.1016/s0091-3057(01)00533-0.

[19] Grimison, P., Mersiades, A., Kirby, A., Lintzeris, N., Morton, R., Haber, P., Olver, I., Walsh, A., McGregor, I., Cheung, Y., Tognela, A., Hahn, C., Briscoe, K., Aghmesheh, M., Fox, P., Abdi, E., Clarke, S., Della-Fiorentina, S., Shannon, J., Gedye, C., Stockler, M. (2020). Oral THC:CBD cannabis extract for refractory chemotherapy-induced nausea and vomiting: A randomised, placebo-controlled, phase II crossover trial. Annals of Oncology: Official Journal of the European Society for Medical Oncology, 31(11), 1553–1560. https://doi.org/10.1016/j.annonc.2020.07.020.

[20] Stith, S.S., Li, X., Orozco, J., Lopez, V., Brockelman, F., Keeling, K., Hall, B., Vigil, J.M. (2022). The effectiveness of common cannabis products for treatment of nausea. Journal of Clinical Gastroenterology, 56(4), 331–338. https://doi.org/10.1097/MCG.0000000000001534.

[21] Chow, R., Basu, A., Kaur, J., Hui, D., Im, J., Prsic, E., Boldt, G., Lock, M., Eng, L., Ng, T.L., Zimmermann, C., Scotte, F. (2025). Efficacy of cannabinoids for the prophylaxis of chemotherapy-induced nausea and vomiting-a systematic review and meta-analysis. Supportive Care in Cancer: Official Journal of the Multinational Association of Supportive Care in Cancer, 33(3), 193. https://doi.org/10.1007/s00520-025-09251-w.

[22] Chang, A.E., Shiling, D.J., Stillman, R.C., Goldberg, N.H., Seipp, C.A., Barofsky, I., Rosenberg, S.A. (1981). A prospective evaluation of delta-9-tetrahydrocannabinol as an antiemetic in patients receiving adriamycin and cytoxan chemotherapy. Cancer, 47(7), 1746–1751. https://doi.org/10.1002/1097-0142(19810401)47:7≤1746::aid-cncr2820470704≥3.0.coS;2-4.

[23] Patti, F., Chisari, C.G., Solaro, C., Benedetti, M.D., Berra, E., Bianco, A., Bruno Bossio, R., Buttari, F., Castelli, L., Cavalla, P., Cerqua, R., Costantino, G., Gasperini, C., Guareschi, A., Ippolito, D., Lanzillo, R., Maniscalco, G.T., Matta, M., Paolicelli, D., Petrucci, L. SA.FE. group. (2020). Effects of THC/CBD oromucosal spray on spasticity-related symptoms in people with multiple sclerosis: Results from a retrospective multicenter study. Neurological Sciences: Official Journal of the Italian Neurological Society and of the Italian Society of Clinical Neurophysiology, 41(10), 2905–2913. https://doi.org/10.1007/s10072-020-04413-6.

[24] Jones, É., Vlachou, S. (2020). A Critical Review of the Role of the Cannabinoid Compounds Δ^9-tetrahydrocannabinol (Δ^9-THC) and Cannabidiol (CBD) and Their Combination in Multiple Sclerosis Treatment. *Molecules. Basel, Switzerland,* 2521 4930https://doi.org/10.3390/molecules25214930

[25] Cannabis-In-Cachexia-Study-Group et al. (2006). Comparison of orally administered cannabis extract and delta-9-tetrahydrocannabinol in treating patients with cancer-related anorexia-cachexia syndrome: A multicenter, phase III, randomized, double-blind, placebo-controlled clinical trial from the Cannabis-In-Cachexia-Study-Group. Journal of Clinical Oncology: Official Journal of the American Society of Clinical Oncology, 24(21), 3394–3400. doi: 10.1200/JCO.2005.05.1847.

[26] Jatoi, A., Windschitl, H.E., Loprinzi, C.L., Sloan, J.A., Dakhil, S.R., Mailliard, J.A., Pundaleeka, S., Kardinal, C.G., Fitch, T.R., Krook, J.E., Novotny, P.J., Christensen, B. (2002). Dronabinol versus megestrol acetate versus combination therapy for cancer-associated anorexia: A North Central Cancer Treatment Group study. Journal of Clinical Oncology: Official Journal of the American Society of Clinical Oncology, 20(2), 567–573. https://doi.org/10.1200/JCO.2002.20.2.567.

[27] Razmovski-Naumovski, V., Luckett, T., Amgarth-Duff, I., Agar, M.R. (2022). Efficacy of medicinal cannabis for appetite-related symptoms in people with cancer: A systematic review. Palliative Medicine, 36(6), 912–927. https://doi.org/10.1177/02692163221083437.

[28] Farokhnia, M., McDiarmid, G.R., Newmeyer, M.N. *et al.* (2020). Effects of oral, smoked, and vaporized cannabis on endocrine pathways related to appetite and metabolism: A randomized, double-blind, placebo-controlled, human laboratory study. Translational Psychiatry, 10(71), https://doi.org/10.1038/s41398-020-0756-3.

[29] Mukhtar, M.S.A., Gupta, R., Balpande, R. (2025). Assessing the neuroprotective benefits of Cannabis sativa in epilepsy management. Brain Disorders, 17, 100180.

[30] Alshaarawy, O., Elbaz, H.A. (2016). Cannabis use and blood pressure levels: United States national health and nutrition examination survey, 2005-2012. Journal of Hypertension, 34(8), 1507–1512. https://doi.org/10.1097/HJH.0000000000000990.

[31] Corroon, J., Grant, I., Bradley, R., Allison, M.A. (2023). Trends in cannabis use, blood pressure, and hypertension in middle-aged adults: findings from NHANES, 2009-2018. American Journal of Hypertension, 36(12), 651–659. https://doi.org/10.1093/ajh/hpad068.

[32] Vallée, A. (2023). Association between cannabis use and blood pressure levels according to comorbidities and socioeconomic status. Scientific Reports, 13(1), 2069. https://doi.org/10.1038/s41598-022-22841-6.

[33] DeFilippis, E.M., Singh, A., Divakaran, S., Gupta, A., Collins, B.L., Biery, D., Qamar, A., Fatima, A., Ramsis, M., Pipilas, D., Rajabi, R., Eng, M., Hainer, J., Klein, J., Januzzi, J.L., Nasir, K., Di Carli, M.F., Bhatt, D.L., Blankstein, R. (2018). Cocaine and marijuana use among young adults with myocardial infarction. Journal of the American College of Cardiology, 71(22), 2540–2551. https://doi.org/10.1016/j.jacc.2018.02.047.

[34] Jeffers, A.M., Glantz, S., Byers, A.L., Keyhani, S. (2024). Association of Cannabis Use With Cardiovascular Outcomes Among US Adults. Journal of the American Heart Association, 13(5), e030178. https://doi.org/10.1161/JAHA.123.030178.

[35] Kamel, I., Mahmoud, A.K., Twayana, A.R., Younes, A.M., Horn, B., Dietzius, H. (2025). Myocardial Infarction and Cardiovascular Risks Associated With Cannabis Use: A Multicenter Retrospective Study. JACC: Advances, 4(5), 101698. https://doi.org/10.1016/j.jacadv.2025.101698.

[36] Peyrot, I., Garsaud, A.M., Saint-Cyr, I., Quitman, O., Sanchez, B., Quist, D. (2007). Cannabis arteritis: A new case report and a review of literature. Journal of the European Academy of Dermatology and Venereology, 21(3), 388–391. https://doi.org/10.1111/j.1468-3083.2006.01947.x.

[37] Colling, M., Souri, Y., Reifsnyder, T. (2024). Tetrahydrocannabinol vape-associated cannabis arteritis in a patient with minimal tobacco exposure. Journal of Vascular Surgery Cases and Innovative Techniques, 11(1), 101673. https://doi.org/10.1016/j.jvscit.2024.101673.

[38] McBane, R.D., Koster, M.J. (2023). Cannabis Arteritis and Recurrent Phlebitis. Mayo Clinic Proceedings, 98(12), 1831–1832. https://doi.org/10.1016/j.mayocp.2023.06.013.

[39] El Omri, N., Eljaoudi, R., Mekouar, F., Jira, M., Sekkach, Y., Amezyane, T., Ghafir, D. (2017). Cannabis arteritis. The Pan African Medical Journal, 26, 53. https://doi.org/10.11604/pamj.2017.26.53.11694.

[40] Hartman, R.L., Huestis, M.A. (2013). Cannabis effects on driving skills. Clinical Chemistry, 59(3), 478–492. https://doi.org/10.1373/clinchem.2012.194381.

[41] Ronen, A., Gershon, P., Drobiner, H., Rabinovich, A., Bar-Hamburger, R., Mechoulam, R., Cassuto, Y., Shinar, D. (2008). Effects of THC on driving performance, physiological state and subjective feelings relative to alcohol. Accident; Analysis and Prevention, 40(3), 926–934. https://doi.org/10.1016/j.aap.2007.10.011.

[42] Weil, A.T. (1970). Adverse reactions to marihuana: Classification and suggested treatment. New England Journal of Medicine, 282(18), 997–1000. https://doi.org/10.1056/NEJM197004302821803.

[43] D'Souza, D.C., Perry, E., MacDougall, L., Ammerman, Y., Cooper, T., Wu, Y.T., Braley, G., Gueorguieva, R., Krystal, J.H. (2004). The psychotomimetic effects of intravenous delta-9-tetrahydrocannabinol in healthy individuals: Implications for psychosis. Neuropsychopharmacology: Official Publication of the American College of Neuropsychopharmacology, 29(8), 1558–1572. https://doi.org/10.1038/sj.npp.1300496.

[44] Minichino, A., Senior, M., Brondino, N., Zhang, S.H., Godwlewska, B.R., Burnet, P.W.J., Cipriani, A., Lennox, B.R. (2019). measuring disturbance of the endocannabinoid system in psychosis: a systematic review and meta-analysis. JAMA Psychiatry, 76(9), 914–923. https://doi.org/10.1001/jamapsychiatry.2019.0970.

[45] Groening, J.M., Denton, E., Parvaiz, R., Brunet, D.L., Von Daniken, A., Shi, Y., Bhattacharyya, S. (2024). A systematic evidence map of the association between cannabis use and psychosis-related outcomes across the psychosis continuum: An umbrella review of systematic reviews and meta-analyses. Psychiatry Research, 331, 115626. https://doi.org/10.1016/j.psychres.2023.115626.

[46] Reid, S., Bhattacharyya, S. (2019). Antipsychotic treatment failure in patients with psychosis and co-morbid cannabis use: A systematic review. Psychiatry Research, 280, 112523. https://doi.org/10.1016/j.psychres.2019.112523.

[47] Marcotte, T.D., Umlauf, A., Grelotti, D.J., Sones, E.G., Sobolesky, P.M., Smith, B.E., Hoffman, M.A., Hubbard, J.A., Severson, J., Huestis, M.A., Grant, I., Fitzgerald, R.L. (2022). Driving Performance and Cannabis Users' Perception of Safety: A Randomized Clinical Trial. JAMA Psychiatry, 79(3), 201–209. https://doi.org/10.1001/jamapsychiatry.2021.4037.

[48] Rogeberg, O., Elvik, R. (2016). The Effects of Cannabis Intoxication on Motor Vehicle Collision Revisited and Revised Addiction. Abingdon, EnglandVol. 1118 1348–1359https://doi.org/10.1111/add.13347

[49] Perisetti, A., Gajendran, M., Dasari, C.S., Bansal, P., Aziz, M., Inamdar, S., Tharian, B., Goyal, H. (2020). Cannabis hyperemesis syndrome: An update on the pathophysiology and management. Annals of Gastroenterology, 33(6), 571–578. https://doi.org/10.20524/aog.2020.0528.

[50] Allen, J.H., De Moore, G.M., Heddle, R., Twartz, J.C. (2004). Cannabinoid hyperemesis: Cyclical hyperemesis in association with chronic cannabis abuse. Gut, 53(11), 1566–1570. https://doi.org/10.1136/gut.2003.036350.

[51]   Rubio-Tapia, A., McCallum, R., Camilleri, M. (2024). AGA clinical practice update on diagnosis and management of cannabinoid hyperemesis syndrome: commentary. Gastroenterology, 166(5), 930–934e1. https://doi.org/10.1053/j.gastro.2024.01.040.

[52]   Simonetto, D.A., Oxentenko, A.S., Herman, M.L., Szostek, J.H. (2012). Cannabinoid hyperemesis: A case series of 98 patients. Mayo Clinic Proceedings, 87(2), 114–119. https://doi.org/10.1016/j.mayocp.2011.10.005.

[53]   Patterson, D.A., Smith, E., Monahan, M., Medvecz, A., Hagerty, B., Krijger, L., Chauhan, A., Walsh, M. (2010). Cannabinoid hyperemesis and compulsive bathing: A case series and paradoxical pathophysiological explanation. Journal of the American Board of Family Medicine, 23(6), 790–793. https://doi.org/10.3122/jabfm.2010.06.100117.

[54]   Young-Wolff, K.C., Adams, S.R., Alexeeff, S.E., Zhu, Y., Chojolan, E., Slama, N.E., Does, M.B., Silver, L.D., Ansley, D., Castellanos, C.L., Avalos, L.A. (2024). Prenatal cannabis use and maternal pregnancy outcomes. JAMA Internal Medicine, 184(9), 1083–1093. https://doi.org/10.1001/jamainternmed.2024.3270.

[55]   Paul, S.E., Hatoum, A.S., Fine, J.D., Johnson, E.C., Hansen, I., Karcher, N.R., Moreau, A.L., Bondy, E., Qu, Y., Carter, E.B., Rogers, C.E., Agrawal, A., Barch, D.M., Bogdan, R. (2021). Associations between prenatal cannabis exposure and childhood outcomes: Results from the ABCD Study. JAMA Psychiatry, 78(1), 64–76. https://doi.org/10.1001/jamapsychiatry.2020.2902.

[56]   Gallagher, T.J., Chung, R.S., Lin, M.E., Kim, I., Kokot, N.C. (2024). Cannabis use and head and neck cancer. JAMA Otolaryngology–Head & Neck Surgery, 150(12), 1068–1075. https://doi.org/10.1001/jamaoto.2024.2419.

[57]   Federal Register [Internet]. Schedules of controlled substances: rescheduling of marijuana [Internet]; [cited 2024 Oct 22]. Available from: https://www.federalregister.gov/documents/2024/05/21/2024-11137/schedules-of-controlled-substances-rescheduling-of-marijuana.

Cheryl J. Fitzer-Attas

Cannabis research: why we will always need more

Introduction

Cannabis has occupied a complex place in the history of medicine, shifting from a widely used therapeutic agent in ancient cultures to a prohibited substance in much of the modern world, and more recently reemerging as a potential treatment option across a range of conditions. Its medical use has been recorded for millennia, with references dating back to traditional Chinese medicine nearly 5,000 years ago, where it was prescribed for ailments such as malaria, rheumatism, and pain [1]. Historical records also document its application in ancient Egypt, Greece, India, and the Middle East for conditions ranging from inflammation to seizures. In the nineteenth century, figures such as William B. O'Shaughnessy, a physician, chemist, and pioneer in pharmacology, played a crucial role in introducing medical cannabis to Western medicine, leading to its inclusion in both the United States and British pharmacopeias and the marketing of tinctures for pain, convulsions, and insomnia [1].

By the early twentieth century, the therapeutic use of cannabis began to decline sharply. The absence of standardized formulations, inconsistent dosing, and the emergence of alternatives such as aspirin and morphine contributed to its removal from pharmacopeias. Increasing concerns about its psychotropic effects, associations with crime, and political factors culminated in prohibition, including the 1937 Marihuana Tax Act in the United States and subsequent classification as a Schedule I substance under the Controlled Substances Act (CSA) of 1970 [2]. These measures not only halted legitimate medical use but also significantly curtailed clinical research for decades.

Reemergence of medical cannabis and the continued need for research

In more recent decades, a growing body of preclinical and clinical studies, alongside patient advocacy and policy changes, reignited interest in cannabis as a therapeutic agent. Indeed, several cannabis-based and cannabis-related medicines have now been approved by regulatory authorities. According to the U.S. Food and Drug Administration (FDA), the agency:

Note: In the development of this chapter, this author has used several LLM products for research and editing support.

has approved one cannabis-derived drug product: Epidiolex (cannabidiol) and three synthetic cannabis-related drug products: Marinol (dronabinol), Syndros (dronabinol), and Cesamet (nabilone). These approved drug products are only available with a prescription from a licensed healthcare provider. Importantly, the FDA has not approved any other cannabis, cannabis-derived, or cannabidiol (CBD) products currently available on the market. [3]

In parallel, other jurisdictions, including Canada, Mexico, and multiple European countries, have authorized products such as nabiximols (Sativex) for specific indications [1].

Despite these regulatory milestones, the acceptance of cannabis in modern clinical practice remains limited and irregular. A persistent lack of differentiation between medical and recreational cannabis use, along with enduring stigma around psychoactive substances, continues to influence physician attitudes, patient perceptions, and public discourse [2]. For many clinicians, the paucity of high-quality randomized controlled trials (RCTs), validated guidelines, or accredited Continuing Medical Education (CME) activities, alongside the mixed conclusions of systematic reviews create uncertainty around efficacy and safety. Consequently, some healthcare providers are reluctant to recommend or prescribe cannabinoid-based medicines. Other providers may trust in the potential benefits of cannabis, yet, out of necessity, are left to prescribe/recommend based on anecdotal evidence and information coming from a variety of grassroots peer networks or non-peer-reviewed sources (grey literature). Patients, in turn, are often guided by minimally trained or educated dispensary staff or pharmacists (depending on geography), resulting in a status quo environment of trial-and-error approaches that lack the safeguards and certainty of evidence-based care.

The gap between widespread public use of cannabis and the slower pace of clinical validation underscores the continued need for rigorous research. In many geographies, patient demand, legislative changes, and commercial market growth have rapidly outpaced the generation of robust scientific evidence. The phrase "we need more research" has become both a political and clinical catchphrase; in the scientific context, it reflects the urgent need for well-designed clinical research that can clarify optimal dosing, safety profiles, therapeutic windows, and patient selection criteria. Bridging these gaps is essential for enabling informed decision-making by healthcare professionals, ensuring patient safety, improving patient outcomes, and supporting the responsible integration of cannabis-based medicines into modern therapeutic practice.

Current state of cannabis and cannabinoid research

Based on a plethora of in vitro and in vivo studies, the complex receptor pharmacology of cannabinoid receptors (CB1 and CB2) has been well elucidated [4]. In fact, the isolation of certain compounds in cannabis (phytocannabinoids), and their physiological receptors initiated a new field of research that led to the discovery of endogenous cannabinoids (endocannabinoids) such as arachidonoyl ethanolamide or anandamide (AEA) and 2-arachidonoylglycerol (2-AG) [4]. Continued research brought to light the existence of additional cannabinoid receptors, GPR55 and TRPV1 [4]. Collectively, mechanistic approaches to studying the cannabinoid system have shown that receptor modulation is associated with central and peripheral physiological responses, which then translate into the therapeutic effects in human pathologies. Sideris et al. [5] describe a neuronal endocannabinoid signaling system that is elegantly and tightly coordinated, effectively functioning as a circuit breaker. Importantly, the endocannabinoid system in each of us uniquely interacts with, and is influenced by, phytocannabinoids and synthetic cannabinoids to create a complex interplay of molecules and individualized responses.

Pharmaceutical development pathway

When aspiring to bring cannabis-derived (and other) medicines to patients, a standardized approach is taken by pharmaceutical and biotechnology companies in the clinical development and regulatory pathways that lead to regulatory approval and commercialization (see Box 11.1). Typically, this begins with small Phase 1 studies to characterize safety, tolerability, and human pharmacology of the compound or biologic. These studies can help define pharmacokinetic properties and oftentimes detect pharmacodynamic or biomarker signals. Phase 2 studies are typically designed to demonstrate proof-of-concept efficacy in the target population, and to select the best dose and/or regimen for future pivotal studies. Protocol entry criteria and endpoints are further refined, and short- to mid-term safety is characterized. Larger Phase 3 trials confirm efficacy and safety for benefit-risk assessments and are designed to support regulatory approval and labeling, as well as health technology assessments for pricing and reimbursement discussions. These trials are typically randomized, double-blind, and placebo-controlled with rigorous prespecified statistical analysis plans in place and are often conducted globally.

Box 11.1: Phases of clinical trials for pharmaceutical clinical drug development
Phase 1: Goals are to characterize safety, tolerability, and human pharmacology (PK/PD) in a first-in-human study; short treatment duration (typically 1–28 days); small number of participants (~20–80), often healthy volunteers.

Phase 2: Goals are to demonstrate proof-of-concept efficacy in the target population, to assess safety, and to select the best dose/regimen for future studies; treatment duration typically several weeks to a few months; larger number of diagnosed trial participants (~100–300).

Phase 3: Goals are to confirm efficacy and safety for benefit-risk assessments and regulatory approval; treatment duration typically several months (or longer when chronic exposure is intended); larger number of participants (typically >300); reasonable duration.

Chemistry, Manufacturing, and Controls (CMC) is a critical aspect of drug development as it ensures every dose that a trial participant (and later a patient) receives is the same medicine, even if drug manufacturers modify process steps, scale up, or change manufacturing sites. It is essential that clinical trial lots are representative of the intended commercial processes, and that critical quality attributes (CQA) such as potency and impurities for small molecules, are identified, controlled, and proven to be stable throughout any manufacturing changes.

For synthetic cannabinoid medications, the CMC and CQA requirements are identical to those applied to conventional small-molecule pharmaceuticals, as their active ingredients are chemically defined and consistently reproducible. Synthetic cannabinoid medications that have been approved in the United States include Marinol® (synthetic THC or dronabinol formulated in soft gelatin capsules) and Syndros® (dronabinol in oral liquid solution), both indicated for anorexia associated with weight loss in patients with AIDS and nausea/vomiting associated with cancer chemotherapy in patients who have not improved with usual antinausea medicines.

Complexities unique to plant-based cannabis products

Plant-derived or extract-based cannabinoid products add CMC/CQA complexity due to multiconstituent profiles, cultivation/extraction-driven batch variability, stability issues, device/formulation challenges, contamination risks, and operational/regulatory controls (Table 1). Reviewing these issues, one can understand why dose standardization, cross-study comparability, and reproducibility are persistent hurdles in cannabis research and why tailored CMC/CQA strategies are foundational for the field. Similarly, cannabis' multicomponent pharmacology, administration-dependent PK, drug interaction considerations, and controlled-substance logistics, introduce additional layers of complexities to clinical development programs (Table 2), beyond those typically seen for small molecule and biologics development. Collectively, these topics explain why dose finding, blinding integrity, endpoint interpretability, and safety signal attribution may require modified and innovative strategies when evaluating cannabis medicines.

Plant-derived cannabinoid medicines with rigorous CMC and clinical evidence have been approved by regulators in several indications. A plant-derived cannabidiol oral solution (Epidiolex®/Epidyolex) is approved and marketed in the United States and Europe for the treatment of seizures associated with the rare epilepsies Lennox–Gastaut syndrome, Dravet syndrome, and tuberous sclerosis complex in patients ≥1 year (US) or >2 years (EU). Plant-derived nabiximols (THC:CBD oromucosal spray, Sativex®) are licensed in the UK and many other countries to treat spasticity (muscle stiffness) in multiple sclerosis patients who have not responded adequately to other treatments; Sativex is not FDA-approved for use in the United States.

Table 1: Relevant CMC/CQA topics and the added complexity introduced with cannabis-derived products.

Category	CMC/CQA topic	Added complexity with cannabis-derived products
Analytical/measurement infrastructure (quality systems)	Reference standards and analytical methods	Limited/expensive certified reference standards for minor cannabinoids/terpenes; need for multianalyte methods
	Specifications and release tests	Setting clinically meaningful specs across multiple actives/impurities; defining acceptance ranges for ratios (e.g., THC:CBD) and terpene fingerprints.
Contamination risks	Contaminants	Risk of pesticides, heavy metals, mycotoxins, microbial toxins, residual solvents, plasticizers/leachables from packaging or vape hardware.
	Microbial quality and water activity	Plant biomass carries high bioburden; spores and environmental microbes; controlling water activity in edibles/oils to prevent growth while preserving volatiles.
	Device–product compatibility (vapes) (also presents device/formulation challenges)	Coil materials, wicking, and operating temperature alter aerosol composition and degradants; leachables/extractables and heavy metals risk.
Cultivation/extraction variability	Botanical raw material control (GACP→GMP)	Agricultural variability requires GACP controls, qualified growers, and tight bridging into GMP to ensure consistent starting material quality.
	Extraction and decarboxylation control	Choice of solvent/supercritical CO_2/terpene recovery affects profile; controlling decarb to target neutral versus acidic species without degrading terpenes.
	Lot consistency	Batch-to-batch chemical variability driven by chemovar genetics, cultivation conditions, harvest timing, drying/curing, and extraction parameters.
	Chemotype profile	A large number of cannabinoids and key terpenes to quantify (major, minor, and acidic forms; THC/CBD ratios; batch drift over time).

Table 1 (continued)

Category	CMC/CQA topic	Added complexity with cannabis-derived products
Device/formulation challenges	Label claim and dose standardization	Potency per unit/serving versus per container; converting mg/mL to mg/serving or mg/puff with tolerances tight enough for clinical use.
	Matrix homogeneity (edibles, oils) and dose delivery (vape, oral, topical)	Achieving uniform dispersion of lipophilic actives in complex matrices; dose-per-unit variability; separating phase in oils; per-puff aerosol delivery variability.
	Equivalence and comparability across forms (also present multiconstituent profile challenges)	Bridging oils/edibles/vapes/topicals is nontrivial due to differing release/absorption and distinct volatile profiles.
Operational/regulatory controls	Traceability and chain of custody	Seed-to-sale tracking, controlled-substance handling, and jurisdiction-specific documentation layered on normal GMP batch records.
Stability	Packaging-product interactions	Terpene sorption into elastomers; oxygen permeability; headspace control; photoprotection requirements for clear versus amber containers.
	Stability of active compounds	Oxidation/isomerization of cannabinoids (e.g., $\Delta 9 \rightarrow \Delta 8$), decarboxylation of acidic precursors (THCA/CBDA), terpene evaporation/loss; light/heat/oxygen sensitivity.
	Storage and distribution	Sensitivity to temperature/light/humidity; terpene off-gassing; need to define excursion limits and in-use stability (opened bottles, vape carts).

CBD, cannabidiol; CBDA, cannabidiolic acid; CMC, Chemistry, Manufacturing, and Controls; CO_2, carbon dioxide; CQA, critical quality attribute(s); THCA, tetrahydrocannabinolic acid; GACP, Good Agricultural and Collection Practices; GMP, Good Manufacturing Practice(s); mg/mL, milligrams per milliliter; THC, $\Delta 9$-tetrahydrocannabinol.

Table 2: Relevant clinical development topics and the added complexity introduced with cannabis-derived products.

Category	Clinical development topic	Added complexity with cannabis-derived products
Biomarkers and endpoints	Biomarkers of target engagement	Few validated, mechanism-specific biomarkers exist for cannabinoid/terpenoid pathways, forcing reliance on exploratory or indirect measures.
	Endpoint strategies	Standard scales and PROs were not designed to capture cannabis-specific effects. Since cannabis products often produce concurrent effects across symptom clusters (e.g., pain relief, less spasticity, better sleep, but possible psychomotor effects), composite endpoints may better capture net clinical benefit (and trade-offs) than any single domain.
	Endpoint adjudication and rescue medication rules	Psychoactivity can influence subjective ratings and affect use of rescue medications. Trials should define objective rescue triggers and permitted actions, use blinded central adjudication to confirm events, and prespecify how postrescue data are analyzed to limit bias and preserve interpretability.
Controlled-substance, regulatory and logistics	Abuse liability and diversion	Controlled-substance status requires enhanced security, chain-of-custody and drug-accountability procedures, and limits on dispensing/take-home quantities; these measures increase site burden and can reduce patient convenience and retention.
	Regulatory and logistics	Additional approvals, import/export permits, and jurisdictional variability in cannabis rules extend start-up timelines and site feasibility. Due to varied regulations worldwide, global studies are particularly challenging.
Delivery and administration	Device use and training (vape/inhaler)	Puff topography, breath-hold, and device temperature control influence delivered dose; standardization and training are trial-critical.
Metabolic drug–drug interactions	Interactions with drug-metabolizing enzymes (CYP450s and UGTs)	Cannabinoids can inhibit or induce major hepatic enzymes, increasing the chance of clinically relevant drug–drug interactions with common medications taken simultaneously.
Pharmacology/PK and exposure	Multicomponent pharmacology	Multiple active constituents (e.g., THC, CBD, minor cannabinoids, and terpenes) act on overlapping pathways, creating additive, synergistic, or opposing effects that complicate MOA, dose-response, and attribution of benefit and risk.

Table 2 (continued)

Category	Clinical development topic	Added complexity with cannabis-derived products
	PK variability by route	Inhalation gives rapid spikes with high between-subject variability; oral shows erratic absorption, strong food effects, and first-pass metabolism; transmucosal/topical add further heterogeneity.
	Model-informed dose selection	Requires integrating PK of multiple active constituents. Ratios of constituents vary across products and over time in vivo, and high lipophilicity leads to adipose uptake and slow release, producing prolonged and variable exposure that complicates modeling and washout.
	Food effects and fed/fasted state (oral)	High-fat meals markedly increase exposure; strict dietary controls and replicable meal content needed.
Trial conduct and blinding	Prior exposure and tolerance	Baseline cannabis use, tolerance, and withdrawal risk affect response and safety; requires stratification and washout strategies.
	Placebo-control and blinding	Psychoactive effects, characteristic odor/taste, and device sensations (e.g., vapor warmth) can unblind participants and staff; active placebos, sensory-matched controls, or terpenoid excipients may be needed.
	Self-titration behaviors	Patients naturally titrate to effect; rigid dosing may underperform real-world use, yet flexible dosing complicates analysis and potential labeling.
	Safety monitoring and signal attribution	Cannabis can produce CNS/psychiatric effects (e.g., anxiety, paranoia; psychosis in susceptible individuals), cardiovascular effects (tachycardia, orthostatic hypotension), and cognitive/psychomotor impairment. Because products are mixtures and delivery devices alter exposure, it is difficult to attribute any adverse event to a specific constituent, ratio, or dose.

CBD, cannabidiol; CNS, central nervous system; CYP450, cytochrome P450; MOA, mechanism of action; PK, pharmacokinetics; PRO, patient-reported outcome; THC, Δ^9-tetrahydrocannabinol; UGT, UDP-glucuronosyltransferase.

Thus, while the basic science of cannabis is quite well-studied [4, 5], one can appreciate from the complexities described in Tables 1 and 2, that translating those mechanistic insights into clinically actionable evidence has been challenging [6, 7]. Accordingly, while the effects of medical cannabis have been studied across various medical indications, results are often mixed. For example, some studies suggest a reduction in opioid use among patients using medical cannabis for intractable pain, indicating potential opioid-sparing effects [8]. However, systematic reviews and meta-analyses have found only a small effect size for cannabis in managing chronic and neuropathic pain, with low confidence in these estimates [9, 10]. Similarly, there are controversial results in the literature regarding the use of medical cannabis for chemotherapy-induced nausea and vomiting, and sleep disorders [11], two indications for which cannabis is widely prescribed.

Toward standardized research methodologies

While many of the challenges for clinical research of cannabis products stem from the inherent heterogeneity of plant metabolites, variability in cultivation and harvesting conditions, and the diversity of product forms (see Box 11.2), administration routes, and dosages, the apparent inconsistencies in clinical data may also reflect differences in study design, patient populations, and outcome measures. Indeed, heterogeneity of protocol elements such as entry criteria, visit frequency, and endpoint selection, is a common issue across pharmaceutical research, often contributing to discrepancies in efficacy and safety findings even for well-characterized synthetic drugs [12].

Box 11.2: Different medical cannabis product forms sold worldwide.
Defining product form is essential because it affects pharmacokinetics, dosing precision, patient experience, and comparability across studies.
- Capsules or soft gels (oral)
- Dried flower (for inhalation or vaporization)
- Edibles and beverages (oral ingestion)
- Oils and tinctures (oral or sublingual)
- Suppositories or sprays (rectal, vaginal, or oromucosal routes)
- Topicals (creams, ointments, patches)
- Vapes or cartridges (inhalation)

These themes highlight the need for more rigorous and standardized research methodologies that can be applied to the study of medical cannabis across different medical conditions. For example, recommendations have been made to more objectively measure cannabis exposure, for example, by using a standard 5-mg THC unit to report the dose of THC administered to a research participant [13, 14]. The authors of this method intentionally based the standard unit on THC content, citing it as the primary pharmacological constituent in cannabis. However, this approach has some important limitations: (1) the effects of a standard THC unit will vary according to the route of administration (e.g., oral vs inhaled); (2) the effects of additional cannabinoids and terpenes, which may negate, complement, or enhance effects of THC, are ignored; and (3) it overlooks individual patient variability in response to THC. Despite these concerns, the use of standardized measures in research is encouraged as they facilitate comparable data collection and analyses, essential for comparison of results across studies.

Moving forward, our research community should also find better ways to close the gap between label claims and actual cannabis chemistry using verified potency testing and batch-level traceability. Researchers need to account for regional differences in product rules and testing standards by capturing such information in study metadata. Adopting common descriptors, reporting checklists, exposure metrics, and research context statements, as promoted by the iCannToolkit and other initiatives [15, 16], will support analyses across sites, states, and countries while still acknowledging the richness of cannabis product diversity.

Legal and regulatory influences on medical cannabis research

Alongside these scientific challenges, the history of cannabis regulation in the United States, with the problematic divergence of federal and state policies, has created a complex legal environment that negatively impacts research. The 1970 CSA classified marijuana (cannabis) as a Schedule I substance, thus imposing strict federal controls that prohibit cultivation, distribution, and possession outside of permitted studies and organizations. As a signatory to United Nations (UN) drug conventions, the United States implements its treaty obligations through the CSA and Drug Enforcement Administration (DEA) systems, which constrain availability of research-grade cannabis through restrictive quota, licensing, and recordkeeping requirements, even for US-only studies.

In direct contrast to US federal law, the last three decades have witnessed the legalization and availability of medical cannabis in more than 80% of US states. For US researchers, this contradiction between federal and state policies introduces ambiguity around compliance and law enforcement, making it more difficult to access products, navigate Institutional Review Board (IRB) processes for ethics approvals, and engage with regulatory bodies [17]. Moreover, as highlighted in a 2024 National Academies consensus report [18], the variability in state rules around product categories, potency caps, testing standards, and approved indications makes it difficult to compare outcomes across states or to confidently assess population-level risks.

For multinational trials, treaty-based controls trigger monitoring by the International Narcotics Control Board for national import/export permits, country-specific scheduling rules, and jurisdiction-specific ethics committee expectations. Both US and international oversight extend study start-up timelines, increase study costs, restrict diversity of study drug, and significantly impede product comparability across geographies. Trial designs are often limited to small, single-site studies with constrained formulations, rather than larger, multicenter RCTs that could produce more generalizable evidence.

Another important consequence of the CSA is that funding for cannabis research from US government and public agencies is minimal and existing grant calls mostly focus on (important, but not sufficient) issues of safety and harm. Overall, therapeutic trials are severely under-resourced, sponsors are deterred, and researchers are forced to rely on limited private and philanthropic sources, constraining the generation of scientific evidence.

Recent progress: rescheduling efforts and regulatory clarity

Researchers can be optimistic from signs that the field is moving in positive directions. Recent reclassification attempts include the U.S. Department of Justice's 2024 and HHS's 2023 proposals to move cannabis (marijuana) from Schedule I to Schedule III [19, 20]. More recent U.S. developments suggest continued movement toward Schedule III treatment for certain medical cannabis products, although the timing and outcome of broader federal rescheduling remain open. Similar rescheduling efforts are noted in the UK, Australia, and Germany [21–23]. FDA's 2016 guidance on botanical drug development [24] and 2024 guidance on cannabis and cannabis-derived compounds [25] help to clarify quality and sourcing expectations and regulatory approaches for clinical research with cannabis products. As regulatory frameworks evolve, there is potential for more standardized products and improved research conditions, which could facilitate better acceptance in medical practice [7]. In parallel, USP/ASTM [26] and the USP Herbal Medicines Compendium [27] have published cannabis quality specifications for study drug characterization that should improve interjurisdiction comparability, regardless of different rescheduling statuses. Moreover, the 2024 National Academies report [18] calls for federal public-health leadership, a national surveillance system, adoption of USP product standards, and expanded NIH/CDC state and tribal research, including studies of how specific state and local regulations affect health and equity.

That report, however, also cautions that without better federal alignment and clear guidance, the fragmented policy landscape will continue to impede rigorous, comparable research needed to guide public health and clinical decision-making. While recent initiatives may streamline certain processes, the operational burdens and overall regulatory environment will likely continue to hold back cannabis research relative to many therapeutic areas [7].

Global regulatory impacts on clinical trial design

Illustrative of this, a recent literature review assessed several countries where medical use of cannabis is permitted to determine how legislation and regulations affect the design of clinical trials in those countries [28]. Current legislative frameworks significantly impacted the conduct and publication of clinical trials by dictating indicated pathologies, production of raw plant material, available products, and prescription and reimbursement processes. Countries with a larger number of publications were the United States, Australia, Israel, and the UK. Although many studies were randomized, and included a placebo group, they often had small sample sizes and included heterogeneous populations (e.g., acute and chronically ill patients; subjects who either previously used or never used cannabis), factors that make it particularly difficult to draw conclusions. In addition, when studies test inflorescences or extracts derived from them, different chemovars are allowed, concentrations are typically provided only for the THC and CBD content (and not for the other active molecules), and diverse administration routes are used, confounding comparisons and interpretations. Often cannabis was administered in addition to other medicines, making evaluation of efficacy or safety even more complex. The authors also note that for some approved conditions under current legislation, such as anxiety, ulcerative colitis, Crohn's disease, and appetite enhancement, data coming out of the trials assessed were not definitive. Of note, the country with the greatest number of allowed disorders (and hence clinical trials) is the Netherlands, most likely due to the longstanding use of cannabis both for medical use and recreational purposes. Italy, on the other hand, faces legal restrictions that hinder clinical trials despite authorization for medical cannabis use, resulting in a lack of specific clinical data and complicating the evaluation of efficacy across different studies. Clearly clinical research with medical cannabis has many obstacles to overcome before researchers can properly design multicenter global (or cross state) trials.

Major gaps in the evidence base: the case for real-world evidence

Given the numerous obstacles and intrinsic complexities for medical cannabis clinical research, one can better understand the reasons behind major gaps in the evidence base for the use of cannabis products, even more so for products that are not manufactured or sold as pharmaceutical grade. For medical cannabis products sold at dispensaries or pharmacies (depending on local regulations), there is a dearth of clinical data related to drug interactions, geriatric use, effects on mental health, long-term safety, and much more, making day-to-day clinical decisions and recommendations difficult. In fact, qualitative research with certain cannabis healthcare practitioners (HCPs) reveals a dedicated community of self-taught healers that rely heavily on peer knowledge and grassroots networks, not medical literature, continuing medical education, or association treatment guidelines [29]. Unfortunately, patient experience with most HCPs often reveals isolated prescribers that have accrued only basic knowledge and may not feel ready or comfortable to answer patients' questions regarding medical cannabis [30].

For both clinical and research contexts, there is a need for reliable and standardized tools to meaningfully assess cannabis use and cannabis-specific effects. A recent review [31] provides a comprehensive survey of 21 commonly cited validated measures, dividing them into three important categories: (1) use frequency or quantity questionnaires ($n = 4$); (2) symptom inventories that assess disordered cannabis use, craving, or withdrawal ($n = 13$); and (3) assessments of use motives, effects, and perceptions ($n = 4$). Notably, most of these measures focus on characterizing usage patterns and abuse concerns. One interesting exception is the 21-item Cannabis Effects Expectancy Questionnaire-Medical (CEEQ-M), which measures beliefs about the effects of cannabis on medical symptoms of pain, insomnia, anxiety, and depression in adults [32]. Indeed, the CEEQ-M is included in the Core Assessment Battery for Cannabis (CABC), a broad set of validated questionnaires used for evaluating the health effects of cannabis by the ongoing Cannabis & Health Research Initiative in their ongoing registry study [33, 34]. However, there still remains a major gap in the availability of clinical outcome assessments and patient-reported outcomes that were specifically conceived, designed, and validated to capture symptomatic effects and other medical benefits that may be unique to medical cannabis treatment.

Despite these gaps in clinical tools and evidence, the consumption of cannabis for medical conditions has exploded in recent years. A 2018 online survey in the United States and Canada reported overall prevalence of self-reported ever cannabis use for medical purposes to be as high as 27% [35]. This points to a large amount of untapped real-world data (RWD) not being collected fast enough. Leveraging real-world evidence (RWE) could address the limitations of RCTs, such as their narrow focus and lack of generalizability, by providing a more comprehensive understanding of cannabis's impact across various conditions and patient groups [36]. Moreover, leveraging RWE helps circumvent some of the regulatory and logistical difficulties, high costs, and extended timelines of medical cannabis RCTs in the face of this wealth of unexploited data.

Untapped potential of real-world evidence

The FDA defines RWD and RWE as follows [37]:
- **"Real-world data** are data relating to patient health status and/or the delivery of health care routinely collected from a variety of sources. Examples of RWD include data derived from electronic health records" [37], wearables and digital health technologies, medical claims data, surveys, and product or disease registries that can inform on health status.
- **"Real-world evidence** is the clinical evidence about the usage and potential benefits or risks of a medical product derived from analysis of RWD" [37].

The Professional Society for Health Economics and Outcomes Research (ISPOR) notes that RWD and RWE can be used to supplement RCTs, which remain the gold standard in assessing a treatment's safety and efficacy [38]. ISPOR also states that the benefit of RWE is in providing insights into the use and outcomes of treatment in everyday life as opposed to the defined and monitored structure of RCTs. RWE captures broader, longer-term, and more representative data [39]. Limitations of RWD/RWE are important to consider, including the quality of data collection and biases due to lack of randomization. RWD/RWE analyses primarily evaluate association, not causation, and should be analyzed and interpreted with the understanding that it cannot provide conclusive evidence of the safety and efficacy of cannabis medicines [40].

Despite its potential to provide insights into the effectiveness and safety of cannabis in diverse patient populations and to provide real-world usage patterns, RWE is underutilized for the study of medical cannabis [36, 41]. The multicomponent nature of cannabis, the siloed product supply, heterogeneous patient populations and use patterns, and fragmented regulatory/distribution landscapes can render conventional trial frameworks ill-suited and insufficient for reliable data collection. Medical cannabis research cannot rely solely on RCTs to guide medical (and wellness) practice and is particularly suited for tapping into the availability and power of RW evidence generation [41]. At lower cost and greater scale, researchers can leverage registries, electronic health records (EHRs), social media, and patient networks, to systematically capture safety signals, dosing practices, cost-effectiveness, quality-of-life, and cannabis-relevant/specific outcomes (see Table 3).

In recent years, the establishment of research consortia and real-world registries are helping to shape high-quality cannabis research and collaboration in this direction. Notable examples are the Florida-based Consortium for Medical Marijuana Clinical Outcomes Research [7], Project Twenty21 (T21) in the UK [42], and The National Cannabis Study supported by Johns Hopkins University and the nonprofit Realm of Caring [34]. Established in June 2019, the Consortium for Medical Marijuana Clinical Outcomes Research currently consists of 11 public and private universities in Florida. Some of their offerings include a grants program, an annual conference, and a research data repository named MEMORY that allows for longitudinal ascertain-

Table 3: Suitability of RWD and RWE for medical cannabis research.

Problem for clinical research with medical cannabis	Suitability for RWD collection with medical cannabis	Call to action for medical cannabis research community
Cannabis has multi-component pharmacology and heterogeneous products	Unlike RCTs, RWD can capture the diversity of products being used in local settings	Medical cannabis products should have clear and interpretable labels providing reliable information on cannabinoid and terpene content
RCT protocols often have strict eligibility requirements (e.g., exclude patients with comorbidities or concomitant medications)	RWE can contextualize controlled cannabis trials by collecting evidence from real-world patient populations	RWE protocols should be rigorously designed using established frameworks and templates
Other parameters of RCTs (duration, setting) are not suited for cannabis-relevant topics such as changing dosing patterns, condition-specific outcomes, or long-term safety data.	Well-designed patient registries can systematically capture these types of data, providing critical infrastructure for cannabis knowledge	Registries that are standardized, prospective, disease/treatment-specific can also be leveraged to recruit into RCTs
Data on medical cannabis usage is not well-reported or coded by physicians in EHRs Standardization and coding (e.g., product type, cannabinoid ratio) remain major gaps	Embedding cannabis use documentation into EHRs could allow structured RWE analyses	Harmonization is needed to define credible RWE streams specific for medical cannabis. Creating standardized fields for cannabinoid type, dose, route is critical for future analyses
Major gap in understanding the cost-effectiveness of cannabis therapies as cannabis most often excluded from insurance claims	Insurance/claims datasets (where applicable) could inform cost-effectiveness of cannabis therapies, though current patchwork legality limits utility compared to pharmaceuticals.	In markets where reimbursement occurs (e.g., Germany, Israel), claims RWE could inform pricing and cost-effectiveness.
Sparse safety protocols	Given cannabis' patient-driven adoption, mining of social media and online patient networks may provide unique early-warning safety/adverse event signals.	Establish validation frameworks to avoid bias when looking for social media signals
Lack of cannabis-specific/relevant clinical scales	Current medical cannabis patients and carepartners across geographies can provide readily accessible cohorts for scale development processes.	Leverage well-established methods to capture patient (and carepartner) input in the development and validation of new scales

Table 3 (continued)

Problem for clinical research with medical cannabis	Suitability for RWD collection with medical cannabis	Call to action for medical cannabis research community
Subjectivity of patient-reported outcomes	Digital health technologies, such as wearables and patient apps can provide large amounts of objective 'big data' for advanced AI/ML analyses	Cannabis research could apply digital biomarkers, wearables, and AI-driven causal inference to stratify responders and emulate trials at lower cost.
Medical cannabis data sources are fragmented (state registries, local dispensaries)	Opportunities exist to better understand cannabis outcomes by combining data across geographies	National and international organizations can develop standards and guidance documents to ensure appropriate methods are used for combining these data

Abbreviations: EHR – electronic health record; RCT- randomized controlled trial; RWD – real-word data; RWE – real-world evidence; AI/ML – artificial intelligence/machine learning

ment of health outcomes [7]. Launched in August 2020, T21 is a multicenter, prospective, observational patient registry accessing clinical records for 2 years of patient follow-up across a variety of conditions. Patients are seen at the same intervals used in standard of care and measurements focus on health-related quality of life, mood and depression, sleep, and condition-specific outcomes [42]. The stated goal of the National Cannabis Study is to measure the impact of medical cannabis use on patient health and determining whether use of cannabis is likely to be helpful, harmful, or result in no change for patients. It primarily focuses on patients who are new to medical cannabis and looking for help with anxiety, pain, cancer, and sleep disturbances [34]. These and other platforms aiming to generate RWE are welcome additions to the research landscape for medical cannabis.

Leveraging modern technology for RWE

To standardize cannabis RWE studies, researchers should align with established RWE frameworks and protocol templates (e.g., FDA/EMA programs, NICE's RWE framework, and the ISPOR/ISPE HARPER protocol) to improve credibility and comparability of evidence packages [43]. RWE approaches should be encouraged by funding and research organizations interested in bridging gaps in clinical knowledge and practice. Modern technologies and analytics (AI, digital health, causal modeling) will be critical to improve RWD capture, analytics, and integration into regulatory/clinical decision-making [39].

The standardization [44], quality, usability [45], and variety, of wearables and other digital health technologies have increased greatly in recent years, offering an efficient and objective methodology for prospective clinical research protocols with medical cannabis "in the wild" [29, 46]. Digital health methodologies, and other RWD sources provide an unparalleled amount of big data for applying machine learning approaches such as causal modeling and heterogeneous treatment effect estimation. Likewise, registries and EHRs offer historical and observational data used in the generation of synthetic control arms. For personalized medicine with cannabis, these advanced analytics have the potential to identify which patients benefit from specific formulations or dosing regimens, find responder subgroups, disentangle product-specific effects, and simulate trial-like conditions [47].

The development of a robust evidence base for medical cannabis will continue to require extraordinary efforts to overcome persistent barriers to clinical research.

RWE provides a pathway to bypass many of these regulatory, methodological, and ethical barriers, allowing stakeholders to leverage both RCT and RWE pathways in building a reliable foundation for future therapeutic use and policy decisions.

Author perspectives

As a clinical scientist and drug developer in the pharmaceutical industry, this author has been trained to judge evidence by its fitness to achieve regulatory approval and to inform dosing, labeling, and clinical practice. With cannabis, research is slowed by supply and scheduling rules, variability in product chemistry, and endpoints that are vulnerable to expectancy effects. In this time of modern technologies, leveraging cannabis heterogeneity and patient variability through RWD collections is both appealing and necessary.

As a patient-reported outcomes (PRO) and digital endpoint developer, this author has learned that measurements are the bridge to reaching reliable conclusions. We need to use PROs that focus on areas where patients report benefits, such as pain interference, spasticity burden, sleep quality, and anxiety symptoms. Digital endpoints will add objectivity, and should be prespecified, version-locked, and validated for context of use. These fit-for-purpose measures, when used consistently across products and routes of administration, will elevate the credibility of RWE claims.

As an entrepreneur, this author envisions that the obstacles to cannabis research become sources of new ideas and business models to support critical infrastructure needs for medical cannabis data collections. For example, innovative solutions for product traceability, assessment toolkits, and cannabis-relevant data collection platforms are sorely needed. This author hopes the call for increased nondilutive funding opportunities and supportive investor networks is heard.

References

[1] Fraguas-Sánchez, A.I., Torres-Suárez, A.I. (2023). Therapeutic Uses of *Cannabis Sativa* L.: Current State and Future Perspectives. In: Preedy, V.R., Patel, V.B., Martin, C.R. (eds). Medicinal Usage of Cannabis and Cannabinoids. Academic Press, 407–445. doi: 10.1016/B978-0-323-89867-6.00010-X.

[2] Arboleda, M.F., Prosk, E. (2021). Barriers for the Prescription of Cannabinoid-based Medicines. In: Narouze, S.N. (ed). Cannabinoids and Pain. Cham, Switzerland: Springer, 145–152. doi: 10.1007/978-3-030-69186-8_20.

[3] U.S. Food and Drug Administration. FDA and Cannabis: Research and Drug Approval Process. Updated February 24, 2023. Accessed August 14, 2025. https://www.fda.gov/news-events/public-health-focus/fda-and-cannabis-research-and-drug-approval-process

[4] Gómez-Cañas, M., Rodríguez-Cueto, C., Satta, V., Hernández-Fisac, I., Navarro, E., Fernández-Ruiz, J. (2023). Endocannabinoid-binding Receptors as Drug Targets. In: Maccarrone, M. (ed). Endocannabinoid Signaling. Methods in Molecular Biology. Vol. 2576, New York, NY: Humana, 67–94. doi: 10.1007/978-1-0716-2728-0_6.

[5] Sideris, A., Lauzadis, J., Kaczocha, M. (2024). The basic science of cannabinoids. Anesth Analg, 138(1), 42–53. doi: 10.1213/ANE.0000000000006472.

[6] Shah, S., Narouze, S. (2024). Cannabis as a therapeutic or snake oil? A desperate call for critical appraisal of the literature. Anesth Analg, 138(1), 2–4. doi: 10.1213/ANE.0000000000006592.

[7] Goodin, A.J., Jyot, J., Cook, R.L., Wang, Y., Hasan, M.M., Winterstein, A.G. (2024). Proceedings of the 2024 Cannabis Clinical Outcomes Research Conference. Med Cannabis Cannabinoids, 7(1), 213–217. doi: 10.1159/000541327.

[8] O'Connell, M., Sandgren, M., Frantzen, L., Bower, E., Erickson, B. (2019). Medical cannabis: Effects on opioid and benzodiazepine requirements for pain control. Annals of Pharmacotherapy, 53(11), 1081–1086. doi: 10.1177/1060028019854221.

[9] Amato, L., Minozzi, S., Mitrova, Z., Parmelli, E., Saulle, R., Cruciani, F., Vecchi, S., Davoli, M. (2017). Revisione sistematica sull'efficacia terapeutica e la sicurezza della cannabis per i pazienti affetti da sclerosi multipla, dolore neuropatico cronico e pazienti oncologici che assumono chemioterapie. Epidemiology and Prevention, 41(5–6), 279–293. doi: 10.19191/EP17.5-6.AD01.069.

[10] Bilbao, A., Spanagel, R. (2022). Medical cannabinoids: A pharmacology-based systematic review and meta-analysis for all relevant medical indications. BMC Medicine, 20, 259. doi: 10.1186/s12916-022-02459-1.

[11] Whiting, P.F., Wolff, R.F., Deshpande, S. et al (2015). Cannabinoids for medical use: A systematic review and meta-analysis. JAMA, 313(24), 2456–2473. doi: 10.1001/jama.2015.6358.

[12] Dickersin, K., Mayo-Wilson, E. (2018). Standards for design and measurement would make clinical research reproducible and usable. Proceedings of the National Academy of Sciences of the United States of America, 115(11), 2590–2594. doi: 10.1073/pnas.1708273114.

[13] Freeman, T.P., Lorenzetti, V. (2020). 'Standard THC units': A proposal to standardize dose across all cannabis products and methods of administration. Addiction, 115(7), 1207–1216. doi: 10.1111/add.14842.

[14] Freeman, T.P., Lorenzetti, V. (2021). A standard THC unit for reporting of health research on cannabis and cannabinoids. The Lancet Psychiatry, 8(11), 944–946. doi: 10.1016/S2215-0366(21)00355-2.

[15] Lorenzetti, V., Hindocha, C., Petrilli, K., Griffiths, P., Brown, J., Castillo-Carniglia, Á. et al (2022). The International Cannabis Toolkit (iCannToolkit): A multidisciplinary expert consensus on minimum standards for measuring cannabis use. Addiction, 117(6), 1510–1517. doi: 10.1111/add.15702.

[16] Cousijn, J., Kuhns, L., Filbey, F., Freeman, T.P., Kroon, E. (2024). Cannabis research in context: The case for measuring and embracing regional similarities and differences. Addiction, 119(9), 1502–1504. doi: 10.1111/add.16460.

[17] Congressional Research Service. The Evolution of Marijuana as a Controlled Substance and the Federal-State Policy Gap. Report No. R44782. Published March 10, 2017. Updated April 7, 2022. Accessed August 23, 2025. https://www.congress.gov/crs-product/R44782

[18] National Academies of Sciences. (2024). Engineering, and Medicine. In: Cannabis Policy Impacts Public Health and Health Equity. Washington, DC: National Academies Press, doi: 10.17226/27766.

[19] Drug Enforcement Administration. (2024). Department of Justice. Schedules of Controlled Substances: Rescheduling of Marijuana; notice of hearing on proposed rulemaking. Federal Register, 89(168), 70148–70149. Published August 29, 2024 https://www.federalregister.gov/docu ments/2024/08/29/2024-19370/schedules-of-controlled-substances-rescheduling-of-marijuana.

[20] U.S. Department of Health and Human Services, Office of the Assistant Secretary for Health. Basis for the Recommendation to Reschedule Marijuana into Schedule III of the Controlled Substances Act. August 29, 2023. Accessed October 12, 2025. https://www.dea.gov/sites/default/files/2024-05/ 2016-17954-HHS.pdf

[21] Home Office. Circular 018/2018: Rescheduling of Cannabis-Based Products for Medicinal Use in Humans. Published November 1, 2018. Accessed October 12, 2025. https://www.gov.uk/govern ment/publications/circular-0182018-rescheduling-of-cannabis-based-products-for-medicinal-use-in-humans/rescheduling-of-cannabis-based-products-for-medicinal-use-in-humans-accessible-version

[22] Therapeutic Goods Administration. Over-the-counter access to low dose cannabidiol. Published December 15, 2020. Accessed October 12, 2025. https://www.tga.gov.au/news/media-releases/over-counter-access-low-dose-cannabidiol

[23] Thiermann, A., Welter-Birk, T., Glockemann, P. Turnaround of medical cannabis in Germany? – Draft bill by new German Government to address alleged missteps. Hogan Lovells. Published July 31, 2025. Accessed October 12, 2025. https://www.hoganlovells.com/en/publications/turnaround-of-medical-cannabis-in-germany-draft-bill-by-new-german-government-to

[24] U.S. Food and Drug Administration, Center for Drug Evaluation and Research. Botanical Drug Development: Guidance for Industry. Published December 2016. Accessed October 12, 2025. https://www.fda.gov/media/93113/download

[25] U.S. Food and Drug Administration. Cannabis and Cannabis-Derived Compounds: Quality Considerations for Clinical Research. Guidance for Industry. Published January 2023. Updated March 28, 2024. Accessed October 12, 2025. https://www.fda.gov/regulatory-information/search-fda -guidance-documents/cannabis-and-cannabis-derived-compounds-quality-considerations-clinical-research-guidance-industry

[26] United States Pharmacopeia; ASTM International. USP–ASTM Cannabis Quality: Potential Actions or Recommendations. Published March 28, 2024. Accessed October 12, 2025. https://www.usp.org/ sites/default/files/usp/document/our-work/DS/uspastm-cannabis-quality-recommendations.pdf

[27] United States Pharmacopeia. Herbal Medicines Compendium Monograph 500608: Cannabis Species Inflorescence. Final Authorized Version 1.0. Published June 16, 2025. Accessed October 12, 2025. https://hmc.usp.org/monograph/cannabis-species-inflorescence-1

[28] Baratta, F., Pignata, I., Ravetto Enri, L., Brusa, P. (2022). Cannabis for medical use: Analysis of recent clinical trials in view of current legislation. Frontiers in Pharmacology, 13, 888903. doi: 10.3389/ fphar.2022.888903.

[29] Fitzer-Attas, C., Joshi, S., Pascal, C. (2022). Evidence-based customer discovery as a first step for developing a decentralized clinical research platform for medical cannabis. Med Cannabis Cannabinoids, 5(1), 36–60. O019 doi: 10.1159/000522395.

[30] Cheng, K.Y.C., Harnett, J.E., Davis, S.R., Eassey, D., Law, S., Smith, L. (2022). Healthcare professionals' perspectives on the use of medicinal cannabis to smanage chronic pain: A systematic search and narrative review. Pain Practice, 22(8), 718–732. doi: 10.1111/papr.13161.

[31] Martin-Willett, R., Elmore, J.S., Phillips, P.X., Bidwell, L.C. (2024). Meaningfully characterizing cannabis use for research and clinical settings: A comprehensive review of existing measures and

proposed future directions. Psychiatry and Clinical Psychopharmacology, 34(1), 82–93. doi: 10.5152/pcp.2024.23645.

[32] Weiss, J.H., Tervo-Clemmens, B., Potter, K.W., Evins, A.E., Gilman, J.M. (2023). The Cannabis Effects Expectancy Questionnaire–Medical (CEEQ-M): Preliminary psychometric properties and longitudinal validation within a clinical trial. Psychological Assessment, 35(8), 659–673. doi: 10.1037/pas0001244.

[33] Dubois, C., Madar, C., Klingensmith, R., Jones, T., Strickland, J., Leoutsakos, J., Thrul, J., Vandrey, R. (2025). Conducting Longitudinal Observational Research on the Effects of Cannabis Use on Health, doi: 10.13140/RG.2.2.15952.21768.

[34] Cannabis and Health Research Initiative. The National Cannabis Study. Accessed October 12, 2025. https://www.cannabisandhealth.org/national-cannabis-study/

[35] Leung, J., Chan, G., Stjepanović, D., Chung, J.Y.C., Hall, W., Hammond, D. (2022). Prevalence and self-reported reasons of cannabis use for medical purposes in USA and Canada. Psychopharmacology (Berl), 239(5), 1509–1519. doi: 10.1007/s00213-021-06047-8.

[36] Schlag, A.K., Zafar, R.R., Lynskey, M.T., Athanasiou-Fragkouli, A., Phillips, L.D., Nutt, D.J. (2022). The value of real world evidence: The case of medical cannabis. Frontiers in Psychiatry, 13, 1027159. doi: 10.3389/fpsyt.2022.1027159.

[37] U.S. Food and Drug Administration. Real-world evidence. Updated September 22, 2025. Accessed October 12, 2025. https://www.fda.gov/science-research/science-and-research-special-topics/real-world-evidence

[38] ISPOR. About real-world evidence. Accessed October 12, 2025. https://www.ispor.org/strategic-initiatives/real-world-evidence/about-real-world-evidence

[39] Parums, D.V. (2025). A review of the importance and relevance of real-world data and real-world evidence. Medical Science Monitor, 31, e951118. doi: 10.12659/MSM.951118.

[40] Graham, M., Lucas, C.J., Schneider, J., Martin, J.H., Hall, W. (2020). Translational hurdles with cannabis medicines. Pharmacoepidemiology and Drug Safety, 29(10), 1325–1330. doi: 10.1002/pds.4999.

[41] Shakeri, A., Zhang, M., Tadrous, M. (2021). The missed opportunity for real-world evidence to shape our understanding of medical cannabis. British Journal of Clinical Pharmacology, 87(3), 732–734. doi: 10.1111/bcp.14550.

[42] Sakal, C., Lynskey, M., Schlag, A.K., Nutt, D.J. (2022). Developing a real-world evidence base for prescribed cannabis in the United Kingdom: Preliminary findings from Project Twenty21. Psychopharmacology (Berl), 239(5), 1147–1155. doi: 10.1007/s00213-021-05855-2.

[43] Sarri, G., Hernandez, L.G. (2024). The maze of real-world evidence frameworks: From a desert to a jungle! An environmental scan and comparison across regulatory and health technology assessment agencies. Journal of Comparative Effectiveness Research, 13(9), e240061. doi: 10.57264/cer-2024-0061.

[44] Goldsack, J.C., Coravos, A., Bakker, J.P. et al (2020). Verification, analytical validation, and clinical validation (V3): The foundation of determining fit-for-purpose for biometric monitoring technologies (BioMeTs).NPJ Digital Medicine, 3, 55. doi: 10.1038/s41746-020-0260-4.

[45] Bakker, J.P., Barge, R., Centra, J., Cobb, B., Cota, C., Guo, C.C. et al (2025). V3+ extends the V3 framework to ensure user-centricity and scalability of sensor-based digital health technologies. NPJ Digital Medicine, 8(1), 51. doi: 10.1038/s41746-024-01322-2.

[46] Platient Research Technologies. Accessed October 12, 2025. https://www.platient.health

[47] Mishra, A., Majumder, A., Kommineni, D., Joseph, C.A., Chowdhury, T., Anumula, S.K. (2025). Role of generative artificial intelligence in personalized medicine: A systematic review. Cureus, 17(4), e82310. doi: 10.7759/cureus.82310.

Mariana Larrea Arias

Cannabis regulation and the global landscape: a focus on Mexico

Introduction

This chapter aims to address issues surrounding the cannabis industry and its regulation, with special attention to the Mexican context. We will analyze how legal definitions become the cornerstone of any effective regulatory framework, explore different regulatory models implemented globally, and examine the challenges Mexico faces in its process of cannabis normalization. Likewise, we will study the most relevant emerging markets worldwide and the lessons that can be drawn from their experiences.

The global cannabis regulatory landscape is experiencing unprecedented transformation, with countries across continents reimagining decades-old prohibition policies. Mexico's journey toward comprehensive cannabis regulation exemplifies both the opportunities and complexities inherent in this transformation. As of 2024, Mexico's cannabis framework remains characterized by a unique legal hybrid: medical cannabis is legal, recreational use is decriminalized but unregulated, and a comprehensive legislative framework remains pending congressional approval [1].

Economic benefits and challenges

Economic expectations have been a significant driver for cannabis policy reform in numerous countries. However, reality has proven more complex than initial projections. The potential economic benefits are undeniable: creation of formal employment, tax collection, investment in research and development, and the transformation of illegal markets into legitimate economic contributors.

Updated market data

The Mexico medical cannabis market reached $555 million in 2024, with projections indicating growth to $1,035 million by 2033, representing a Compound Annual Growth Rate (CAGR) of 7.2% [2]. These figures reflect the current medical-only market and do not include potential recreational market revenues.

In Mexico, with an illegal market estimated in billions of dollars annually, regulation would represent an extraordinary opportunity for economic formalization. While comprehensive market projections vary significantly depending on regulatory framework assumptions, the medical cannabis sector alone demonstrates substantial growth potential, driven by progressive legalization, regulatory advancements, and rising patient demand for alternative treatments [2]. Mexico's favorable climate and lower production costs compared to the United States and Canada position it strategically as both a domestic market and a potential export hub within Latin America.

However, international experience reveals considerable challenges in tax policy and banking access. Balancing revenue collection with the need to maintain competitive prices against the black market has proven complex, almost impossible in some jurisdictions. Canada initially established high tax rates which, combined with rigorous regulatory restrictions, allowed the illegal market to maintain a significant portion of total consumption.

The cannabis industry faces uniquely burdensome tax structures that derive directly from lingering stigma. In the United States, Section 280E of the Internal Revenue Code prohibits businesses involved in "trafficking" Schedule I substances from deducting ordinary business expenses [3], resulting in effective tax rates as high as 70–90% for legal cannabis operations. This punitive tax treatment – unparalleled in any other legal industry – stems directly from Nixon-era drug policies specifically designed to target minority communities and political opponents, as later admitted by Nixon's own domestic policy advisor John Ehrlichman [4].

Access to financial services represents another crucial obstacle. Even in legalized markets, banking institutions frequently refuse to serve cannabis companies due to concerns about reputational risks and international regulatory compliance. In Mexico, where banking is dominated by multinational institutions, the context is different because cannabis is not federally illegal or criminalized at the federal level; therefore, medical cannabis businesses or businesses related with industrial uses such as cosmetics, food, beverages, and dietary supplements can open accounts and manage funds [1]. Having worked with cannabis businesses across different jurisdictions, I have observed how this financial exclusion forces companies into cash-only operations in some markets, creating security risks and limiting growth potential.

Regulatory compliance costs constitute both the key to a successful business but also a significant barrier to entry for small producers and entrepreneurs. In Canada, security, traceability, and quality control requirements have favored large corporations with sufficient capital to implement the necessary systems. Mexico must carefully consider how to balance the need for rigorous standards with the importance of creating an inclusive market that benefits communities traditionally involved in cultivation.

An additional challenge for Mexico will be determining how cannabis regulation will interact with international trade agreements, particularly the United States-Mexico-Canada Agreement [5]. Regulatory discrepancies with key trading partners could limit export opportunities and complicate cross-border supply chains, though as legal experts note, the treaty's provisions do not explicitly address cannabis trade, leaving significant uncertainty about future cross-border commerce [6].

Cannabis regulation in Mexico: current landscape

Historical challenges and the current state of cannabis regulation

The process of cannabis regulation in Mexico has followed a particularly winding path, driven more by judicial rulings than by legislative initiative. The first amparo in Mexican cannabis history was actually a medical amparo promoted by the Grace Elizalde family in 2015, whose 8-year-old daughter suffered from severe epilepsy {7}. This case established crucial precedents for medical necessity and constitutional rights arguments, with Grace becoming the first person to legally import a cannabis-derived substance for medical purposes in Mexico in August 2015 [8]. This groundbreaking medical case became the catalyst for subsequent legal developments and directly influenced the 2017 legislative reforms.

The journey continued significantly in November 2015, when the Supreme Court of Justice (SCJN) granted an amparo to four members of the Mexican Society for Responsible and Tolerant Self-Consumption, establishing for the first time that the absolute prohibition of cannabis for recreational purposes violated the right to free development of personality [9]. In this 4–1 ruling, the court found that prohibitions on using marijuana violated the "right to the free development of personality" [10].

The Grace Elizalde family's medical amparo case provided the legal foundation that directly led to the decisive milestone of 2017 when Congress approved reforms to the General Health Law to allow medicinal and industrial uses of cannabis, albeit limiting tetrahydrocannabinol (THC) content to 1%. This legislative modification represented a direct response to the constitutional arguments established in the Grace Elizalde case and other medical necessity precedents. While this represented a significant advance in medical access, it left pending the regulation of adult (recreational) use, maintaining the recreational market outside the authorized medical framework.

Judicial pressure continued to accumulate, and in 2018, after the issuance of the fifth amparo on the recreational matter, the SCJN declared five articles of the General Health Law that prohibited the sowing, cultivation, harvesting, preparation, and acquisition of cannabis unconstitutional. This accumulation of jurisprudence led the Court to formally request the Legislative Branch to modify the unconstitutional articles, establishing a deadline that, after multiple extensions, expired without Congress fulfilling its obligation [11].

In response to this legislative inaction, in June 2021, the Supreme Court issued a general declaration of unconstitutionality, effectively eliminating the criminalization of personal cannabis consumption [12]. However, this ruling left a significant regulatory vacuum, as it decriminalized consumption without establishing a framework for legal production and distribution, creating a paradox where it is legal to consume but not to acquire or produce cannabis commercially.

Implementation issues

The implementation of cannabis regulation in Mexico has faced significant obstacles that reflect both political resistance and technical complexities. Despite the SCJN's declaration of unconstitutionality, health authorities have shown reluctance to issue the self-consumption permits that the same ruling contemplates. This institutional resistance has created a legal limbo where consumers have a recognized right but lack practical mechanisms to legally exercise it. Applications for personal use permits soared 62.5% in 2024, reaching 8,967 in 10 months, far outpacing 2023's 5,516 applications, reflecting growing public demand despite regulatory ambiguity [13].

In the realm of medicinal and industrial cannabis, the slow publication of the corresponding regulations (which took almost 4 years from the 2017 legal reform) and the restrictive interpretations of COFEPRIS (Federal Commission for Protection against Sanitary Risks) have hindered market development. Pioneer companies have had to resort to judicial amparos to obtain authorizations that, in principle, should be granted within the existing regulatory framework.

Verified case study

An emblematic case is that of Xebra Brands, which received court-ordered authorizations from COFEPRIS through an amparo process in 2022 to develop a business of hemp-derived cannabidiol (CBD) products, from cultivation to the final product [14]. This precedent illustrates both the existing legal possibilities and the practical obstacles that companies must overcome to operate within the current legal framework.

These implementation challenges have generated a fragmented and confusing market: on one hand, CBD products with less than 1% THC circulate increasingly openly in formal commerce; on the other hand, recreational consumption operates in a gray zone where users have constitutional protection but lack legal access to regulated and safe products.

While Mexico's implementation process has been gradual, this developmental stage represents an opportunity to learn from international experiences and build a more comprehensive framework. Mexico is one of the unique countries in the world with a full pharmaceutical medical cannabis regulation; re-scheduling cannabis since 2017; developing industrial cannabis businesses focused on the wellness industry; also working on the development of new industrial products for the construction and textile industry; decriminalizing cannabis since 2021 and developing legal figures for cannabis club spaces where consumers are allowed to be educated while using their consumption permits in a safe environment. The country has already established important legal foundations through the 2017 General Health Law reforms and subsequent Supreme Court rulings, creating a platform for future growth. Rather than rushing into implementation, Mexico can leverage this extended process to develop innovative approaches that address both social justice concerns and market development, potentially creating a more inclusive and sustainable cannabis industry than models implemented elsewhere.

Pilot programs: key access pathways for medical cannabis

Before analyzing pilot programs, it is crucial to establish a fundamental conceptual distinction that has been overlooked in industry discourse: the difference between *medical cannabis* and *pharmaceutical cannabis*. This differentiation explains much of the developmental stagnation in countries like Mexico, despite having comprehensive legal and regulatory frameworks favoring cannabis. Medical cannabis encompasses two distinct industries: flower dispensaries and magistral formulations (as exemplified by Switzerland, Canada, and Uruguay), operating under medical supervision but with flexible preparation standards. In contrast, pharmaceutical cannabis operates under full Good Manufacturing Practices (GMP) compliance, requiring expensive licenses, sanitary authorizations, and Big Pharma-level regulatory infrastructure that current cannabis companies cannot financially sustain.

This conceptual confusion becomes particularly evident when examining countries like Germany, which lacks pharmaceutical legislation specific to cannabis products and restricts access to flower through medical dispensaries while remaining closed to magistral formulations. How can meaningful technological and scientific development in pharmaceutical cannabis products occur without appropriate laws and industry infrastructure to promote such advancement? Consequently, Mexico has emerged as the gateway country for European and German companies, serving as host and holder of sanitary registrations developed abroad, driven by two key factors: its substantial patient market and its role as the access point to Latin America. Understanding this regulatory and conceptual distinction between medical cannabis and pharmaceutical cannabis is essential for addressing the industry's developmental challenges and designing appropriate regulatory pathways.

In my legal practice, I have observed that one of the most promising yet underutilized pathways for medical cannabis access lies in magistral formulations – personalized medications prepared by pharmacists according to individual prescriptions. Unlike standardized pharmaceutical cannabis products, magistral formulations allow healthcare providers to customize cannabinoid ratios, concentrations, and delivery methods to meet specific patient needs. This approach represents a fundamental shift from the industrial pharmaceutical model, where patients must adapt to available products, to a patient-centered model where medications are tailored to individual therapeutic requirements.

Switzerland has emerged as a pioneer in this approach through its groundbreaking pilot program launched in 2022, which allows participating patients to access cannabis through magistral preparations under medical supervision [15]. This program, involving multiple cities including Zurich and Basel, serves as a crucial proof-of-concept for understanding patient populations and optimizing regulatory frameworks. The Swiss model demonstrates how controlled access programs can generate invaluable data about dosing patterns, patient demographics, therapeutic outcomes, and potential risks, providing evidence-based foundations for broader policy decisions. Unlike theoretical regulatory frameworks, these pilot programs reveal the practical realities of medical cannabis use in real-world clinical settings.

The accessibility advantages of magistral formulations cannot be overstated, particularly in emerging markets. Where standardized pharmaceutical cannabis products may cost patients \$800–1,200 monthly, magistral preparations can often be produced at a fraction of this cost while offering superior therapeutic customization. Pharmacists can adjust CBD-to-THC ratios, incorporate minor cannabinoids like cannabigerol (CBG) or cannabinol (CBN), and select, and select optimal delivery methods based on patient-specific factors such as metabolism, condition severity, and tolerance levels. This personalization is particularly crucial for pediatric patients, older adult populations, and individuals with complex medical conditions who may not respond adequately to standardized formulations.

The cost barriers associated with pharmaceutical cannabis products represent a fundamental obstacle to equitable access, yet emerging solutions offer promising pathways toward democratization. Having worked with patients across different jurisdictions, I have witnessed both the challenges of prohibitive pricing and the success of innovative access models. Magistral formulations, social cultivation programs, and public health integration represent viable strategies for overcoming cost barriers. Countries implementing insurance coverage, government production programs, and pharmacy compounding have successfully expanded access beyond affluent populations. Mexico's robust pharmaceutical infrastructure and tradition of social medicine position it well to implement these alternative approaches, potentially creating one of the world's most accessible medical cannabis systems.

Mexico exemplifies the paradox of comprehensive regulatory frameworks coupled with inadequate practical access. Despite having legal provisions for medical cannabis since 2017, the country lacks a robust infrastructure for product availability and affordability. COFEPRIS has authorized several pharmaceutical cannabis products, but their prohibitive costs and limited availability through pharmacies create insurmountable barriers for most patients. This situation highlights how regulatory compliance without accessibility considerations can result in technically legal but practically inaccessible medical cannabis systems. The magistral formulation pathway offers a potential solution by leveraging Mexico's established pharmaceutical compounding infrastructure while providing cost-effective access to personalized cannabis medications.

Countries considering cannabis regulation should implement pilot programs as essential policy development tools. These programs provide several critical benefits: they generate real-world data about patient populations and consumption patterns, identify potential public health impacts, test regulatory mechanisms before full implementation, and build public and medical professional confidence through controlled, monitored access. Countries that proceed directly to full legalization without pilot programs often encounter unexpected challenges that could have been identified and addressed through smaller-scale experimentation. For Mexico, implementing pilot programs in select states or cities could provide invaluable insights for optimizing the national regulatory framework while ensuring patient access during the extended legislative process, and to understand who are the patients.

Legal strategies for semisynthetic hemp compounds and import navigation

In my practice representing cannabis industry clients, I have developed specific strategies for navigating the regulatory landscape surrounding semisynthetic hemp compounds and importation procedures. These approaches have proven effective in providing clients with legally compliant pathways to market entry while minimizing regulatory risk.

For semisynthetic hemp compounds, the key strategy involves establishing clear documentation of the hemp source material and the transformation process. Mexican customs and health authorities focus primarily on the final product's THC content, the final use, and the source plant's legal status. Therefore, maintaining comprehensive chain-of-custody documentation from certified hemp cultivation through synthesis to final product testing creates a defensible legal position. Additionally, obtaining pre-clearance letters from COFEPRIS regarding product classification can prevent costly delays and regulatory challenges.

Import strategies must address both customs procedures and health authority requirements. I recommend clients establish relationships with specialized customs brokers experienced in cannabis-adjacent products and maintain detailed technical dossiers for each product line. Working proactively with COFEPRIS to obtain health registrations before importation attempts significantly improves success rates. For clients dealing with novel cannabinoids, engaging in pre-consultation processes with regulatory authorities helps establish precedents and clarify classification procedures for future market entrants.

The Mexican context becomes particularly relevant considering that the country implemented intensive security strategies to address organized crime networks. Since 2006, while these policies succeeded in dismantling major cartel leadership structures, they also contributed to more than 350,000 homicides and approximately 100,000 disappearances. Importantly, today's security challenges have evolved beyond cannabis trafficking, with synthetic drugs like fentanyl representing the primary contemporary threat to public health and security. Effective cannabis regulation would represent not only a policy refinement but a strategic reallocation of enforcement resources toward more pressing contemporary drug threats.

From an international perspective, Mexico has the potential to become a principal producer of legal cannabis, taking advantage of its ideal climatic conditions, centuries of agricultural experience, and competitive production costs. With an adequate regulatory framework, the country could position itself as a significant exporter of medicinal cannabis and hemp-derived products, adding value to its traditional agricultural production.

Additionally, Mexican regulation would represent a fundamental counterweight to the prohibitionist policies that the United States has historically imposed in the region. Together with advances in Uruguay and Canada, a Mexico with regulated cannabis would consolidate a critical mass of American countries with viable alternative policies, facilitating a rethinking of international drug control treaties that currently hinder global reform.

The Mexican experience offers invaluable insights for cannabis regulation in complex institutional contexts, positioning the country as a potential leader in developing innovative regulatory approaches. While Mexico faces unique challenges including organized crime presence and institutional complexity, these circumstances have also fostered creative legal and policy solutions that may prove more applicable to other developing nations than models developed in more privileged contexts. Mexico's approach – combining constitutional jurisprudence, community participation, and gradual implementation – could establish a new paradigm for cannabis regulation that balances pragmatic governance with social justice considerations, offering a blueprint for countries facing similar structural challenges while securing Mexico's competitive advantages in this transformative sector.

Global markets to watch: Brazil, Turkey, Australia, Thailand, and Germany

Brazil: push for decriminalization and the emerging medical cannabis sector

Brazil, Latin America's largest economy and home to more than 200 million people, has historically maintained a conservative stance toward cannabis. However, in recent years, it has experienced significant advances, particularly in both medicinal and industrial hemp sectors. In 2015, the National Health Surveillance Agency (ANVISA) allowed, for the first time, the importation of CBD-based medications for specific conditions. The regulatory framework has significantly expanded since then.

Recent regulatory breakthrough

A watershed development occurred in November 2024 when Brazil's Superior Court of Justice (STJ) issued a landmark ruling legalizing hemp with up to 0.3% THC for medicinal and industrial purposes [16]. The STJ's ruling established that hemp with THC content below 0.3% does not fall under the restrictions of Brazil's Narcotics Law, as it lacks psychoactive properties, and mandated ANVISA to regulate hemp cultivation and commercialization by May 2025 [17].

The Brazilian medicinal cannabis market has shown exponential growth, expanding from fewer than 1,000 import authorizations in 2015 to serving approximately 672,000 patients across more than 80% of Brazilian municipalities by 2024. Importantly, the 2024 hemp law has begun to address the cost barriers that previously limited access to privileged socioeconomic sectors. The legislation allows domestic production of CBD-rich hemp varieties specifically for pharmaceutical processing, which is expected to significantly reduce product costs by eliminating import dependencies. ANVISA has also established new pathways for magistral cannabis preparations, enabling pharmacies to compound personalized cannabis medicines at substantially lower costs than imported pharmaceutical products.

In parallel, the Federal Supreme Court has been considering the decriminalization of cannabis possession for personal consumption, with recent 2024 developments suggesting increasing judicial momentum toward reform. The Court's discussions have been influenced by the successful implementation of the industrial hemp framework, which has demonstrated Brazil's capacity for effective cannabis regulation. Additionally, several Brazilian states have implemented medical cannabis cultivation pilot programs under federal oversight, creating precedents for expanded domestic production capabilities.

Brazil's phased approach – beginning with medical imports, expanding to domestic hemp cultivation, and developing cost-effective magistral preparations – offers a pragmatic roadmap for expanding access across socioeconomic sectors. The Brazilian model particularly shows how industrial hemp legalization can serve as a stepping stone to broader cannabis policy reform while building public and institutional confidence in regulated cannabis systems.

Turkey: emerging medical cannabis hub and regional leadership

Turkey has emerged as a surprising leader in medical cannabis regulation within the Middle East and Europe bridge region, implementing comprehensive reforms in 2024 that position the country as a significant player in the global medical cannabis sector. The Turkish Ministry of Agriculture and Forestry approved large-scale cannabis cultivation licenses for pharmaceutical purposes, with initial plantings covering over 1,000 hectares across multiple provinces, including Antalya, Izmir, and Samsun. This regulatory framework allows both Turkish and international pharmaceutical companies to establish cultivation and processing facilities under strict government oversight.

The Turkish approach prioritizes pharmaceutical-grade production with GMP compliance from cultivation through processing. The government has established the Turkish Medicines and Medical Devices Agency (TITCK) as the primary regulatory body, implementing streamlined licensing procedures that have attracted substantial foreign investment.

Turkey's strategic geographic position has enabled it to target both European and Middle Eastern markets for medical cannabis exports. The country's existing pharmaceutical manufacturing infrastructure and established trade relationships provide significant advantages for scaling cannabis medicine production. Recent agreements with European health agencies have facilitated export pathways, while domestic medical cannabis prescriptions have begun for conditions including epilepsy, chronic pain, and cancer-related symptoms, with products available through hospital pharmacies.

Australia: medical cannabis maturity and future recreational potential

Australia represents one of the most successful examples of implementing a comprehensive medicinal cannabis system. Since federal legalization in 2016, the country has developed a sophisticated regulatory framework under the supervision of the Office of Drug Control and the Therapeutic Goods Administration. This system allows prescriptions for a wide range of medical conditions, without limiting itself to a restrictive list of ailments, leaving the final decision to professional medical judgment.

Updated patient statistics

The Australian medicinal cannabis market has grown exponentially, expanding from a few dozen patients in 2017 to more than 338,000 active patients by late 2024, with estimates suggesting the total user base may reach 700,000–900,000 when including all cannabis medicine users [18].

A notable feature has been the rapid evolution of medical perceptions: initially skeptical, the Australian medical community has significantly increased prescription rates as clinical evidence and practical experience have demonstrated the utility of cannabis for various conditions.

Australia has also developed a robust local production industry, complemented with strategic imports, creating an ecosystem where large corporations, specialized small producers, and research institutions coexist. The country has positioned medicinal cannabis as an emerging export sector, leveraging its international reputation for high-quality standards in pharmaceutical and agricultural products.

In the recreational realm, the Australian Capital Territory (which includes Canberra) legalized possession and personal cultivation in small quantities in 2020, creating a precedent that other Australian states are considering following.

Thailand: regulatory reversal and policy recalibration

Thailand's cannabis journey represents one of the most dramatic policy reversals in global cannabis regulation history. After initially becoming the first Southeast Asian country to legalize medicinal cannabis in 2018, Thailand took the unprecedented step of completely decriminalizing cannabis in 2022, removing it from controlled narcotics lists.

Critical policy update

In 2024, Thailand's government initiated moves to re-criminalize recreational cannabis use, representing a major reversal of its 2022 decriminalization policy [19]. Prime Minister Srettha Thavisin announced in May 2024 plans to have cannabis re-listed as a narcotic by year's end [20]. As of late 2024, the regulatory framework maintains medical cannabis access through licensed clinics and hospitals while prohibiting recreational sales and public consumption, with penalties reinstated for non-medical use.

The initial Thai approach prioritized local economic development, with government promotion of cannabis as a commercial crop for traditional farmers and policies protecting the industry from foreign corporate domination. However, the rapid proliferation of cannabis dispensaries and cafes following decriminalization, coupled with concerns about youth access and tourism-related issues, prompted the 2024 regulatory reversal.

The regulatory correction has significantly impacted Thailand's cannabis tourism industry, which had briefly transformed Bangkok and tourist destinations into cannabis-friendly zones. Many dispensaries and cannabis cafes were forced to close or pivot to medical-only operations under strict licensing requirements. The government has implemented a transition period allowing existing operators to obtain medical cannabis licenses, but enforcement has become increasingly strict, with raids on establishments selling cannabis for recreational purposes.

A unique aspect of the Thai model has been its integration with traditional medicine. Unlike other countries that have prioritized Western pharmaceutical approaches, Thailand has formally recognized and promoted the use of cannabis in traditional Thai medicine preparations, validating ancestral knowledge.

For Mexico, Thailand's experience provides crucial lessons about the importance of comprehensive regulatory planning before implementation. The Thai reversal demonstrates how rapid liberalization without adequate regulatory frameworks can lead to political backlash and policy corrections that disrupt established markets. Thailand's current medical-only approach, however, shows how traditional medicine integration and local producer protection can be maintained within more restrictive frameworks, offering valuable insights for Mexico's own regulatory development process.

Germany: Europe's largest cannabis market, medical expansion, and recreational implementation

Germany currently represents Europe's largest and most sophisticated medicinal cannabis market. Since medicinal legalization in 2017, the country has developed a comprehensive system where cannabis products are prescribed by doctors and reimbursed by the public health insurance system for various conditions, eliminating economic barriers to access for insured patients.

Updated market figures

Medical cannabis sales in Germany exceeded 420 million euros (approximately USD 455 million) in 2024, with projections indicating the market could reach 1 billion euros by 2028 [21]. Patient numbers rose from approximately 200,000–250,000 in early 2024 to approximately 338,000 by late 2024, with estimates suggesting the total user base including all cannabis medicine consumers may reach 700,000–900,000 [22].

A significant development in 2024 has been expanded patient access to cannabis flower products, with insurance reimbursement now covering a broader range of cannabis flower varieties and preparations, moving beyond the previous emphasis on standardized pharmaceutical extracts.

The German model is characterized by high pharmaceutical quality standards. All products must comply with European GMP, establishing a standard that has influenced quality requirements internationally. However, these requirements sometimes necessitate processing that affects plant quality. Initially dependent on imports, mainly from Canada and the Netherlands, Germany has gradually developed domestic production capacity through a rigorous bidding system.

In 2021, the German coalition government announced plans to legalize cannabis for adult use, a project that culminated in April 2024 with the approval of a law that allows personal consumption and limited domestic cultivation, with plans to establish non-profit "cannabis clubs" for collective production and distribution. However, the implementation has faced significant challenges.

Recent regulatory complications

German health authorities proposed amendments in 2024 to restrict telemedicine prescriptions for medical cannabis, requiring mandatory in-person doctor consultations before cannabis prescriptions can be issued [22]. These amendments would also ban mail-order medical cannabis dispensing and impose stricter rules for medical cannabis prescriptions generally [23].

These regulatory developments have created complications for both the cannabis club model and medical patients. Members seeking legal access through cannabis clubs must navigate complex medical consultation requirements, potentially limiting the accessibility that the law intended to provide. The telemedicine restrictions have also affected medical cannabis patients, with some insurance providers requiring additional in-person consultations for cannabis prescriptions, potentially creating access barriers despite the system's universal coverage principles.

For Mexico, the German experience highlights both the benefits of insurance-integrated medical cannabis systems and the potential complications of overly restrictive consultation requirements that could inadvertently limit access for legitimate patients. The lesson is clear: while quality standards and medical oversight are essential, regulatory frameworks must balance rigor with practical accessibility to avoid creating unnecessary barriers to patient care.

Comparative analysis: how these markets differ and what can be learned from them

The comparative analysis of these five markets reveals significant patterns and contrasts that offer valuable lessons for Mexico. First, we observe that regardless of the cultural or political starting point, the global trend is toward some form of normalization and regulation, albeit with diverse rhythms and scopes.

Implementation models show significant variations: while Australia and Germany have prioritized highly regulated medical approaches with pharmaceutical standards, Thailand initially opted for broad liberalization before reverting to medical-only frameworks in 2024. Turkey has emerged as a pharmaceutical-focused model targeting export markets, while Brazil has transitioned from import-dependent systems to domestic hemp cultivation with expanded access mechanisms following its November 2024 legal breakthrough.

A key differential factor has been economic accessibility. Systems that include health insurance coverage (Germany) or domestic production capabilities (Brazil's new hemp framework, Turkey's pharmaceutical production) have significantly democratized access, while models based exclusively on commercial imports have historically resulted in prohibitive prices for the majority of the population.

The participation of small producers and traditional communities also varies enormously: while Thailand initially implemented policies to protect local farmers before its regulatory reversal, Turkey has emphasized pharmaceutical partnerships, and Australia has maintained corporate consolidation trends. Brazil's 2024 hemp law attempts to balance farmer participation with technical standards. This tension between social inclusion and technical standards represents one of the greatest regulatory challenges across all markets.

For Mexico, these experiences suggest the importance of designing a regulatory model that responds to national socioeconomic realities, instead of simply importing frameworks developed in more privileged contexts. The fundamental lesson from these diverse experiences – including Thailand's reversal, Brazil's hemp breakthrough, Turkey's rapid implementation, and Germany's telemedicine challenges – is that successful cannabis regulation requires careful planning, phased implementation, and continuous adaptation based on real-world outcomes rather than theoretical frameworks.

Conclusion: the path forward

The global cannabis landscape is at a historic moment of transformation. The trend toward normalization and regulation is clear and irreversible, although specific models vary considerably. This regulatory diversity represents both a challenge and an opportunity: on one hand, it complicates international trade and standardization; on the other hand, it allows observing and learning from multiple approaches in real time.

The future of cannabis regulation in Mexico stands at a critical crossroads. After years of advances driven primarily by the judicial branch, the country now faces the challenge of building a comprehensive regulatory framework that addresses the complexities of all aspects of cannabis: medicinal, industrial, and adult use. This process occurs in a changing political context, where the upcoming years could mean either acceleration or stagnation of the regulatory process.

In the immediate horizon, Mexico needs to develop better regulatory and commercial alliances with countries where regulations are homologized, such as Canada and Brazil, to establish an international market that allows all countries to have better understanding. Also needed is a shift in regulatory focus from business models to patient access, in order to reach real patients effectively.

The medicinal cannabis sector, although technically legal since 2017, urgently requires more comprehensive development of its regulations to facilitate patient access and allow the flourishing of a national industry that is currently severely limited by restrictive interpretations from the health authority.

The potential cannabis market in Mexico is extraordinary: with more than 127 million inhabitants, a privileged geographic location, and ideal agricultural conditions, the country could develop a multibillion-dollar industry spanning from medicinal products to innovative industrial applications. The medical cannabis sector alone is projected to grow from USD 555 million in 2024 to over USD 1 billion by 2033, demonstrating substantial economic potential even within current regulatory constraints. However, materializing this broader potential requires overcoming regulatory uncertainties that currently discourage significant investments.

International collaboration will be fundamental in this process. Mexico can benefit enormously from the regulatory experiences of countries such as Canada and Australia, adapting them to its specific context. Particularly valuable will be exchanges with other Latin American countries such as Colombia and Brazil, which have managed to develop successful approaches despite sharing some similar sociopolitical challenges.

The fundamental lesson that emerges both from the Mexican experience and the global landscape is that effective cannabis regulation does not merely constitute a technical exercise but a strategic opportunity that requires carefully balancing market development with public health considerations. The regulatory structures Mexico builds today will determine not only the development of a domestic industry but also the country's position in the emerging global cannabis economy, offering a blueprint for countries facing similar implementation challenges while securing Mexico's competitive advantages in this transformative sector.

References

[1]   CMS Law (2024). Cannabis law and legislation in Mexico. CMS Expert Guides. Retrieved from: https://cms.law/en/int/expert-guides/cms-expert-guide-to-a-legal-roadmap-to-cannabis/mexico

[2]   IMARC Group (2024). Mexico Medical Cannabis Market Size, Share, Report 2033. Retrieved from: https://www.imarcgroup.com/mexico-medical-cannabis-market

[3]   Internal Revenue Service. (1982). Internal Revenue Code Section 280E. United States Treasury Department.

[4]   Baum, D. (2016, April). Legalize it all: How to Win the War on Drugs. Harper's Magazine, https://harpers.org/archive/2016/04/legalize-it-all/.

[5]   U.S. Customs and Border Protection. (2025). U.S.–Mexico–Canada Agreement (USMCA). U.S. Department of Homeland Security, https://www.cbp.gov/trade/priority-issues/trade-agreements/free-trade-agreements/USMCA/FAQs.

[6]   Harris Sliwoski (2024). USMCA and Cannabis Trade. Retrieved from: https://harris-sliwoski.com/cannalawblog/usmca-and-cannabis-trade/

[7]   Shirole, T. (2015, September 7). 8-year-old Mexican Girl Paves Way to Using Medical Cannabis Legally. MedIndia, https://www.medindia.net/news/8-year-old-mexican-girl-paves-way-to-using-medical-cannabis-legally-153200-1.htm.

[8]   El País (2015, September 8). Grace, the girl who changed the marijuana law in Mexico. Retrieved from: https://elpais.com/internacional/2015/08/17/actualidad/1439832269_141655.html

[9]   The New York Times (2015, November 5). Ruling in Mexico sets into motion legal Marijuana. Retrieved from: https://www.nytimes.com/2015/11/05/world/americas/mexico-supreme-court-marijuana-ruling.html

[10] The Guardian (2015, November 4). Mexico supreme court rules ban on marijuana use unconstitutional. Retrieved from: https://www.theguardian.com/world/2015/nov/04/mexico-supreme-court-recreational-marijuana-legal

[11] Foley, Lardner, L.L.P. (2021, July 2). Recreational Use of Cannabis Moves Forward in Mexico by Decision of the Supreme Court. JDSupra, https://www.jdsupra.com/legalnews/recreational-use-of-cannabis-moves-3812515/.

[12] NORML (2021, June 29). Mexico: Supreme Court Moves to Abolish Laws Prohibiting Personal Use of Marijuana. Retrieved from: https://norml.org/blog/2021/06/29/mexico-supreme-court-moves-to-abolish-laws-prohibiting-personal-use-of-marijuana/

[13] Rio Times Online (2024). Mexico's Marijuana standoff: Permits tighten as use surges. Retrieved from: https://www.riotimesonline.com/mexicos-marijuana-standoff-permits-tighten-as-use-surges/

[14] Xebra Brands (2022). Xebra Completes Last Legal Requirement to Receive Mexican Cannabis Authorizations. Retrieved from: https://xebrabrands.com/en/xebra-completes-last-legal-requirement-to-receive-mexican-cannabis-authorizations/

[15] Swiss Federal Office of Public Health. (2022). Cannabis Pilot Program: Implementation and Early Results. Bern: Federal Department of Home Affairs.

[16] Licks Legal (2024, November). Brazilian Superior Court of Justice authorizes cultivation of industrial hemp for medical purposes. Retrieved from: https://www.lickslegal.com/news/brazilian-superior-court-of-justice-authorizes-cultivation-of-industrial-hemp-for-medical-purposes-and-sets-regulation-deadline-for-anvisa

[17] Souto Correa Advogados. (2025, February 20). Hemp regulation in Brazil: STJ reiterates Anvisa's obligation to regulate the topic by May 2025. https://www.soutocorrea.com.br/en/artigos/hemp-regulation-in-brazil-stj-reiterates-anvisas-obligation-to-regulate-the-topic-by-may-2025/

[18] Australian Therapeutic Goods Administration. (2024). Medicinal Cannabis Data Report. Canberra: Australian Government Department of Health.

[19] CNN (2024, January 10). Thailand moves to ban recreational cannabis use, 18 months after decriminalizing it. Retrieved from: https://www.cnn.com/2024/01/10/asia/thailand-cannabis-reverse-proposed-laws-intl-hnk

[20] Holpuch, A. (2024). Thailand Prime Minister Seeks to Criminalize Weed in Reversal. New York Times Company. https://www.nytimes.com/2024/05/08/world/asia/thailand-cannabis-narcotic-law.html

[21] MJBizDaily (2024). Report: German medical cannabis sales to exceed $455M in 2024. Retrieved from: https://mjbizdaily.com/german-medical-cannabis-sales-to-exceed-455m-in-2024-report-says/

[22] Forbes (2025*). Germany Is Seeking Stricter Rules For Medical Cannabis To Crack Down On Online Prescriptions. Retrieved from: https://www.forbes.com/sites/dariosabaghi/2025/10/06/germany-is-seeking-stricter-rules-for-medical-cannabis-to-crack-down-on-online-prescriptions/

[23] Business of Cannabis (2025*). Germany Advances MedCanG Amendments to Restrict Prescriptions and Telemedicine Use. Retrieved from: https://businessofcannabis.com/germany-advances-medcang-amendments-to-restrict-prescriptions-and-telemedicine-use/

Index